Digital Audio Workstation

Colby Leider

McGraw-Hill

New York   Chicago   San Francisco
Lisbon   London   Madrid   Mexico City
Milan   New Delhi   San Juan   Seoul
Singapore   Sydney   Toronto

The McGraw-Hill Companies

Cataloging-in-Publication Data is on file with the Library of Congress.

1 2 3 4 5 6 7 8 9 0 DOC/DOC 0 1 0 9 8 7 6 5 4

ISBN 0-07-142286-2

The sponsoring editor for this book was Steve Chapman and the production supervisor was Pamela A. Pelton. It was set in New Century Schoolbook by Patricia Wallenburg. The art director for the cover was Anthony Landi.

Printed and bound by RR Donnelley.

McGraw-Hill books are available at special quantity discounts to use as premiums and sales promotions, or for use in corporate training programs. For more information, please write to the Director of Special Sales, McGraw-Hill Professional, Two Penn Plaza, New York, NY 10121-2298. Or contact your local bookstore.

 This book is printed on recycled, acid-free paper containing a minimum of 50% recycled, de-inked fiber.

For Kristine

Contents

Contents vii

Contents

Introduction

In the late 1970s and 1980s, digital recording and editing was a very expensive proposition. In 1978, $115,000 would buy a 32-track digital multitrack recorder; ten years later, a full-blown 48-track digital multitrack recorder/editor ran about one-quarter of a million dollars. The three-year-old laptop on which I write these words, coupled with a single audio interface, can record and mix about 96 tracks of digital audio for a total cost of under $4000. And it can do all of that while simultaneously running a bunch of audio-processing plug-ins that cost a very small fraction of their hardware outboard-gear counterparts. The computer-based digital audio workstation has reached a level of maturity—in sound quality, technical capability, and sheer value—that now commands the attention of musicians from all walks of life and all budgets. That which just recently required a dedicated hardware mixing console, a myriad of audio gear, frequent maintenance, and a huge monetary investment can now be quite efficiently accomplished within a single general-purpose computer by outfitting it with a handful of off-the-shelf audio components to create a streamlined audio-processing workhorse.

This book is about that workhorse, the digital audio workstation (DAW), and it has been written for a wide range of audiences. My hope is that it will be useful for general readers and beginning students alike—for both the technically minded and the not-so-technically minded—in learning about the nature of sound, the basics of digital audio, how to set up a digital audio workstation, and a tour of the available technologies that make the DAW a robust audio production tool. The book interweaves musical, technical, and aesthetic considerations with the broader goal of instilling the inseparable importance of all these components to the successful use of the digital audio workstation as more than just a tool, but in fact a musical instrument.

The history of computers in all forms of music-making is wonder-fully complex and rich, full of stories of creative hacks, cobbled-together gear, and wires hanging out everywhere. Whenever I think of the days of punch-card computing, painfully slow processors, and 10 Mb disk drives, I marvel that any music was made with them at all. (I would surely not have had the patience or perseverance!) But some really great music was made with those computers with such limited technical capabilities, and that very fact in many ways paved the road for the modern digital audio workstation—along with the modern technical conveniences like the ready and inexpensive avail-ability of very fast computers, very large storage devices, and a vari-ety of great software! But without the pioneering and innovative use of computers in the world of music, the modern digital audio worksta-tion might have arrived much later, even given a plethora of modern, fast computers.

This book is not a manual meant to accompany a particular soft-ware program, but instead it attempts to dissect some of the larger issues involved in making music on modern computers. Because the field of computers and music—as with all fields relying heavily upon technology—lies in a state of perpetual flux, anything more than a cursory effort devoted to explaining the specifics of current hardware and software seems like an exercise in futility. After having lived through a couple of cycles of software and hardware coming and going, I have always found it far more useful and engaging to explore and teach critical concepts common to virtually all ways we can make music with computers. With that in mind, I hope that the present vol-ume piques some aspect of your curiosity—whether it lies in making music, recording music, writing music software, inventing new hard-ware, or some combination of these—that will excite you to explore further this remarkably exciting and challenging field.

Acknowledgments

Thank you to all of my family, friends, and colleagues who supported my writing of this book. Thank you particularly to Steve Chapman, my editor at McGraw-Hill, for his help, support, and confidence, and to Ken Pohlmann for his mentorship and indefatigable guidance. To Julius Smith for reading a portion of the draft, thank you. To Kristine H. Burns for helping especially with pedagogical aspects of the book, thank you. To Paul Koonce for insightful discussions and extended phone calls, thank you. To Joe Abatti for technical assistance, thank you. To Alan Shockley for frequent pep talks, thank you.

And to Kristine and Liam who let me work in the evenings and over the weekends, thank you.

Colby Leider
Homestead, Florida

About the Author

Colby N. Leider is a noted composer, inventor of audio control equipment, instrument builder, and one of the world's leading authorities on digital audio workstations. With degrees in electroacoustic music and music composition from Princeton and Dartmouth, he currently teaches at the University of Miami, Florida, as an Assistant Professor of Music Engineering. Leider also holds a bachelor's degree in electrical engineering from the University of Texas at Austin. A leading figure in digital music, he has presented compositions throughout the world, written numerous articles and papers, and presented research on a variety of digital audio topics at leading professional gatherings.

Computers, Sound, and Digital Audio

Operating Systems

Most people have at least a general understanding of what a computer is, how it works, and what it can do, but few of us ever stop to really contemplate the notion of the computer operating system. We are excited when a new version of the computer operating system we currently use arrives, but truly new paradigms for interacting with computers are much more rare. With the proliferation of microcontrollers (inexpensive, miniature computers, of which the typical car has as many as 50 or 60, for example), even the most basic machines we use in our daily lives now include an operating system. From home thermostats to cell phones, burglar alarms to car stereos, and refrigerators to home computers, operating systems are increasingly dictating the language we use when communicating with machines.

More generally, an operating system provides an interface between human intention and the vast resources and capabilities of the computer. At the most basic level, interfaces provide a vocabulary—graphical, textual, auditory, haptic, or some combination of these—with which we can construct a syntax to tell machines what we want them to do. When we work with computers, we know we can concatenate the basic "words" of an operating system in different ways to create a complex suite of commands. The ability to construct a complex command syntax is not limited to computers now; consider the typical home-theater remote control which can be programmed to send macros, or a "sentence" of commands, to audiovisual equipment. As operating systems become more robust and user-friendly, individual commands like "power on" and "volume up" have been superseded by high-level grammatical constructions. On a modern remote control, a single button or voice command for "watch the evening news" can turn on the television, tune the appropriate channel, activate the audio system at the desired volume level, dim the lights, and close the draperies.

In the digital audio world, a kind of friendly schism exists between those who exclusively mix sound on dedicated hardware mixing consoles and those who do the same exclusively on general-purpose computers. This schism has been bridged somewhat in recent years by hybrid systems in which a so-called "control surface," or gestural interface that controls computer-based mixing software using a traditional mixer form factor, enables mix engineers to leverage their existing abilities and training in using the mixer to control software-based mixers. The line has also been blurred by "intelligent consoles," or mixing boards driven by a built-in operating system and custom-made host computer. On a traditional analog mixing console, for example, a dedicated knob might control some aspect of equalization, like the cutoff frequency of a low-pass filter; on an "intelligent console," the functionality of the knob is programmed by the mix engineer in the console's operating system. (Incidentally, the concept of the "programmable knob" in digital music systems dates back to the Synclavier, the first digital sound synthesizer, invented in the 1970s.)

However, programmability presents advantages and disadvantages. In the traditional analog mixing console, a certain comfort and facility is achieved from knowing that a particular knob always controls a particular filter, for example. However, programmability allows the engineer to configure the physical layout of the mixer in any conceivable way.

In any case, hardware mixing consoles are predicated on a particular interface as well, just as the general-purpose computer is. The features of the interface (whether knobs, buttons, and faders, or a mouse, keyboard, and software) dictate how we interact with the machine to create music. Whether working with a mixing console or a digital audio workstation in a modern recording studio, we must be aware that the operating system resident on each dictates the way we communicate with the system. Knowing as much as possible about the interface with which we work is tantamount to working efficiently and effectively. When creativity strikes, it would be frustrating to have to reach for the operating manual!

Unsound Sound: Sound Unwound

Before delving into the digital audio workstation, which we have not yet formally defined, we should pause to discuss sound itself, jumping back a few millennia. Remarkably early in Western history, Aristotle (384–322 B.C.) discussed the physical nature of sound and that a medium of some kind is required for its transmission:

> Actual sound requires for its occurrence two such bodies and a space between them; for it is generated by an impact.... Further, we must remark that sound is heard both in air and in water, though less distinctly in the latter. Yet neither air nor water is the principal cause of sound. What is required for the production of sound is an impact of two solids against one another and against the air. The latter condition is satisfied when the air impinged upon does not retreat before the blow, i.e., is not dissipated by it.

That is why it must be struck with a sudden sharp blow, if it is to sound—the movement of the whip must outrun the dispersion of the air, just as one might get in a stroke at a heap or whirl of sand as it was traveling rapidly past. (*De Anima* II.8)

This hypothesis of sound as organized variations in air pressure was the prevailing sentiment for many centuries, later echoed and developed by Boethius. The wave nature of sound was challenged by Pierre Gassendi (1592–1655), who formulated an atomic theory of sound, an outgrowth of a philosophical movement of the time known as atomism. He incorrectly thought that sound was the organized, rhythmic motion of atomic sound particles through a medium such as air. (Incidentally, Gassendi, who was a contemporary of Galileo, also made the first attempt to measure the speed of sound.)

In 1686, Isaac Newton reasserted Aristotle's air-motion theory of sound, developing a mathematical theory of sound propagation as a wave phenomenon in his classic work *Principia Mathematica*. Figure 1.1 illustrates a page from Newton's text, in which he describes how sound waves incident upon a small pinhole in a wall create a new point-source of acoustic wave propagation. We now understand well that sound is simply alternating waves of high and low air pressure ("compressions" and "rarefactions") that, upon hitting our ears, cause our brain to invoke some kind of auditory sensation using one or several of various neural mechanisms. Consider waves washing ashore at the beach; the peaks exhibit a higher amount of water pressure relative to the troughs. The peak pounds us with water pressure when we get in its way; the trough pulls us back out into the ocean. Let's replace these high and low waves of water pressure by high and low waves of air pressure. Voila! We have sound. We don't see the waves anymore (because they are simply comprised of air), so the analogy breaks down there. (Incidentally, Boethius in the sixth century compared sound waves to ocean waves, so I can't take credit for the analogy!)

As an aside, however, we can in fact see sound under some circumstances. It was discovered, in 1934, that some materials can exhibit a property known as *sonoluminescence*: in the presence of pulsating sound waves, tiny gas bubbles of a certain chemical composition can "light up" with each pulse of the sound wave. (See Figure 1.2.) How and why this happens is still not fully understood, even though Keanu Reaves and Morgan Freeman so keenly and thoroughly investigate it in the 1996 Hollywood film Chain Reaction.

Figure 1.1
A page from Sir
Isaac Newton's
Principia (1686).

PRINCIPIA MATHEMATICA. 359

PROPOSITIO XLII. THEOREMA XXXIII.

Motus omnis per fluidum propagatus divergit a recto tramite in spatia immota.

Caf. 1. Propagetur motus a puncto *A* per foramen *BC*, pergatque, fi fieri poteſt, in ſpatio conico *BCQP*, ſecundum lineas rectas divergentes a puncto *A*. Et ponamus primo quod motus iſte ſit undarum in ſuperficie ſtagnantis aquæ. Sintque *de, fg, hi, kl*, &c. undarum ſingularum partes altiſſimæ, vallibus totidem inter-

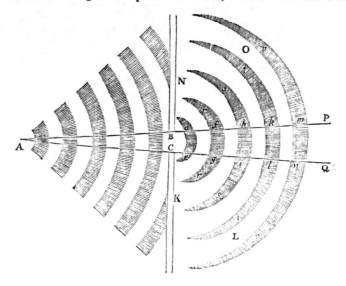

mediis ab invicem diſtinctæ. Igitur quoniam aqua in undarum jugis altior eſt quam in fluidi partibus immotis *LK. NO*, defluet eadem de jugorum terminis *e, g, i, l*, &c. *d, f, h, k*, &c. hinc inde verſus *KL & NO*: & quoniam in undarum vallibus depreſſior eſt quam in fluidi partibus immotis *KL, NO*; defluet eadem de parti-

Sound does not need air to travel, but just a medium of some kind; anything but a perfect vacuum will do fine. We can make an inexpensive telephone by connecting two tin cans with a long length of string, because the string itself can conduct sound pressure waves. (Actually, the air pressure of our voices speaking into one can is transduced into mechanical vibrations of the string and then back to air pressure at the other tin can. We will talk more about this in Chapter 8.)

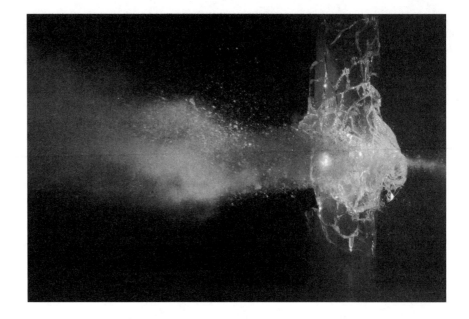

It might also help to create an analogy to sound propagation with a simple mechanical spring. If one places the spring on a table and pushes (compresses) one end, the compression travels through the spring. Think of these compressions and rarefactions along the spring as a visual model for compressions and rarefactions of air pressure. Notice another analogy to sound waves with the spring: Once the compression reaches the end of the spring, it travels back, just as sound waves can "bounce" off of walls. (Figure 1.3).

Warning: If equations and mildly technical prose make you fall asleep or drool, feel free to skip ahead a few pages. The equations aren't complicated at all, though, and the mildly technical prose is rather interesting, if I do say so myself. (And I do.)

As mentioned, sound waves need a medium of some kind in which to travel. Sound cannot travel in a vacuum. Dolphins have known that sound travels even better underwater than it does in air for quite some time. The composer Michel Redolfi (b. 1951) has even produced entire underwater concerts of his music. Why does sound travel better underwater? Sound travels faster underwater because there is less acoustical resistance. Two simple equations describe the speed of sound in any medium, be it air, water, string, or the adjoining wall in my old apartment, which my next door neighbors enjoyed blasting with their subwoofers (yes, plural) late at night, every night:

$$v_g = \sqrt{\gamma R T} \qquad (1)$$

$$v_s = \sqrt{\frac{E}{\rho}} \qquad (2)$$

Figure 1.3
(a) A simple tin-can telephone; (b) wave propagation through a Slinky; and (c) amount of compression in the Slinky as a function of position at a given point in time.

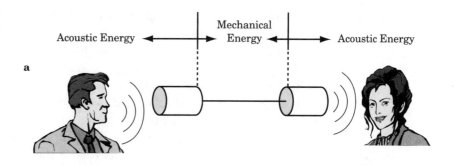

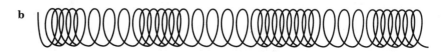

The first equation describes the speed of sound in a gas. Here, the Greek letter gamma (γ) is a constant called the *adiabatic index* (generally taken to be 1.4 for air) that relates to the specific heat of the medium; R represents the gas constant (taken as 287 J/(kg•K) for air); and T is the absolute temperature of the gas in Kelvins at 1 atmosphere (atm) of air pressure. Plugging these numbers into Equation 1, we calculate the speed of sound at standard temperature and pressure (STP)[1] air is around 340 meters per second, or about 750 miles per hour.

---

[1] STP is defined to be 273 Kelvins (0 degrees Celsius) at 1 atmosphere (760.0 millimeters of mercury, or 101,325 pascals) of pressure.

The second equation describes the speed of sound in a solid. In this equation, *E* is the *Young's modulus* of the medium (a measure of the stiffness of the solid), and the Greek letter rho (ρ) is the density of the solid. The values of both *E* and ρ can be quickly found by looking them up in a table or by searching the Internet. Note that the speed of sound is fastest in solids, slower in liquids, and even slower in gases (such as air).

Incidentally, the ratio of the speed of an object in a particular medium to the inherent speed of sound of that medium is called the object's *Mach number*. An object traveling at Mach 1 is thus traveling at the speed of sound. Because this number is a ratio of two numbers with identical units, it is dimensionless. Of course, as we've seen, the actual speed that Mach number denotes varies according to temperature and pressure, for example.

It is possible to derive other equivalent equations to describe the speed of sound based on other properties of the transmission medium, such as atmospheric pressure, elevation above sea level, humidity, density of a solid, and so on. As a point of curiosity, Table 1.1 gives some idea of the speed of sound in various media. A simple formula that approximates the speed of sound in air as a function of temperature is

$$V = 331 + (0.6 \times T) \tag{3}$$

where T is the temperature in degrees Celsius.

TABLE 1.1

The Speed of Sound in Various Gases and Solids at 293.15 K and 1 Atmosphere of Pressure

Element	Approximate Speed of Sound (m/sec)
Sodium	3200
Potassium	2000
Gold	1740
Silver	2600
Platinum	2680
Argon	319
Nitrogen	334
Oxygen	317
Hydrogen	1270
Iron	4910

Decomposing Sound in Frequency and Time

Tuning forks are labeled according to the fundamental frequency at which they vibrate when struck. For example, striking a 440 Hz tuning fork causes it to vibrate back and forth 440 times per second. (The unit of Hertz, named after the German physicist Heinrich R. Hertz [1857–1894], is a measure of frequency formerly called *cps*, or *cycles per second*.)

Ideally, the tuning fork only vibrates at 440 Hz, but in reality it also tends to vibrate at integer multiples of 440 Hz, particularly when we strike it harder. When it vibrates only at 440 Hz, we call the sonic result a *pure tone* because it consists of only one frequency (a single sine wave, as illustrated in Figure 1.4). When multiples of this fundamental frequency are present (for example, when the fork is struck sufficiently harder), we call the sonic result a *complex tone*. These

Figure 1.4
A moving tuning fork with a pencil attached can literally generate a graphical depiction of a sine wave. Note the period and wavelength of the resulting motion.

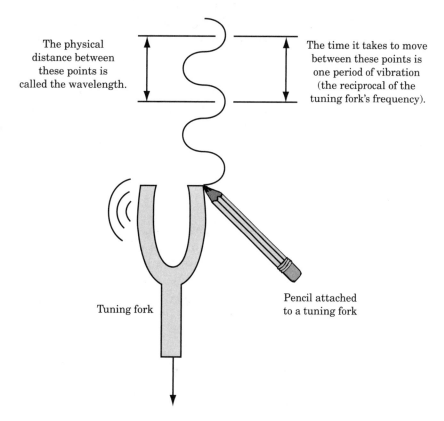

The physical distance between these points is called the wavelength.

The time it takes to move between these points is one period of vibration (the reciprocal of the tuning fork's frequency).

Tuning fork

Pencil attached to a tuning fork

multiples are called *harmonics* or *overtones* if they are integer multiples of the fundamental frequency. If they are not integer multiples of the fundamental, we usually just call them *partials*. In fact, many objects in real life exhibit *periodicity*, which simply means that they do something (for example, vibrate) in a repetitive (periodic) manner: tuning forks, swings, merry-go-rounds, and economic indicators represent just a few.

We say that the tuning fork that vibrates 440 times per second has a *frequency* of 440 Hz, and the *period* of each vibration is 1/440 second; the frequency and period of a waveform are reciprocals of each other. Said another way,

$$f = \frac{1}{T} \tag{4}$$

The period is the time it takes for the wave to propagate through one complete cycle of air pressure rarefaction and compression.

Another term used to describe sound waves is the *wavelength*. As the name implies, the wavelength of a periodic sound is the physical length of one complete cycle of rarefaction and compression. Wavelength must be related to both the frequency and speed of a wave's propagation, but how? Well, the speed of sound has units of meters per second (m/s), and frequency has units of Hz (or 1/s). If we divide meters per second by the reciprocal of seconds, the units of seconds will cancel, resulting in meters, as we want. So we can say that a pure tone's wavelength, λ, is related to its speed of propagation and its frequency by

$$\lambda = \frac{v}{f} \tag{5}$$

In the early nineteenth century, Jean-Baptiste Fourier postulated that, under a certain set of conditions, signals of any complexity can be represented as an infinite sum of pure-tone "building blocks." The pure tones are, as we just saw, simply sine waves (or more generally referred to as sinusoids). Think about that: Virtually any signal (no matter if it represents audio data, video data, the closing value of a stock index over a period of time, or the daily high temperature for the past month) can be decomposed into sinusoidal components.

Fourier's Theorem tells us mathematically how to determine what sinusoidal frequencies combine to make a complex signal. It also shows us how to compute their relative strengths (*amplitudes*) and how they should align in time (*phases*). A two-dimensional graph that displays the amplitudes of a a signal's constituent sinusoidal components is loosely referred to as the signal's *spectrum*.

Figure 1.5 illustrates the spectrum of a short clarinet sound. The x axis represents frequency in Hertz. The y axis displays relative amplitudes of each of the clarinet's sinusoidal components; the darker the color, the greater the amplitude of that frequenccy.

Figure 1.5

The spectrum of a short sample of a clarinet.

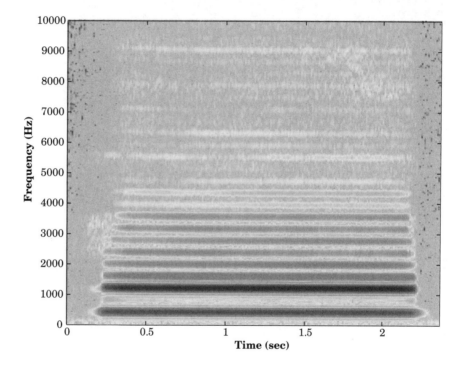

This process can also work in reverse: We can build complex signals from sinusoids. To illustrate this, let's graphically build a 440-Hz square wave out of sine waves. (See Figure 1.6.) Notice that the more sine waves we add, the closer the result looks to a perfect square wave. Fourier's Theorem, the mathematical explanation of which is beyond the scope of our discussion here, tells us which frequencies to add together to yield a desired result. In this instance, it would tell us to combine integer multiples of a fundamental frequency, omitting the even multiples. Thus, if we combine our fundamental (1×440 Hz) with 3×440Hz, 5×440 Hz, 7×440Hz, and so on ad infinitum, we will have produced a perfect square wave. The more of these *over-*

Figure 1.6
Building a square
wave from
sinusoids. (a)
Fundamental
only; (b) with the
third (odd)
harmonic; (c) with
harmonics 3 and
5; (d) with
harmonics 3, 5,
and 7; (e) with
harmonics 3, 5, 7,
and 9; and (f)
with the first
1,000 odd
harmonics.

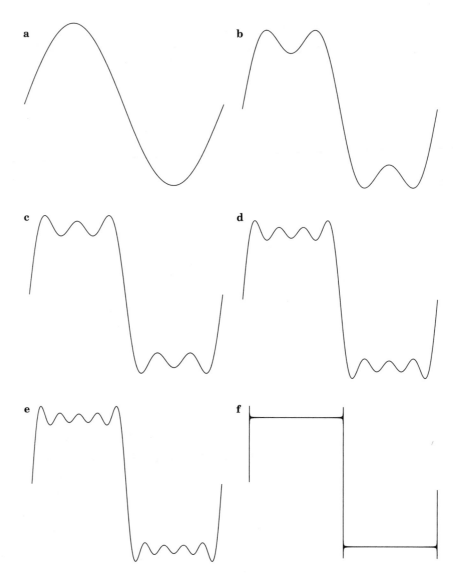

tones we include in our summation, the closer we get to our desired result.

We often describe sounds as "bright" or "dark." Words like these generally refer to higher concentrations of energy in the high or low frequency regions, respectively, of the sound's spectrum. We can manipulate the spectrum of audio in different ways. One such way is through the use of filters that boost (amplify) or cut (attenuate) a portion of a sound's spectrum. To filter a signal and change its spectral content, an audio signal is simply passed through the filter, and what emerges from this operation is the filtered audio signal.

Filters generally assume one of four characteristic forms: lowpass, highpass, bandpass, or band-reject (notch). (See Figure 1.7.)

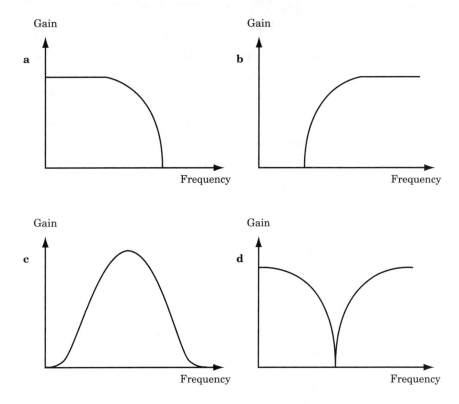

Figure 1.7
Idealized shapes of (a) low-pass filters; (b) bandpass-filters; (c) band-reject filters; (d) high-pass filters.

Lowpass filters allow only frequencies that fall below a specified threshold to pass. In the range of frequencies that the filter passes, called the *passband*, the filter has a gain of unity (i.e., it passes the input signal with a gain of 1). The cutoff threshold is often called the −3 dB point,[2] because it is defined as the frequency at which the reduction amount is three decibels (or about 70%) of the gain in the passband region. (See Figure 1.7.)

Conversely, highpass filters pass only frequencies above cutoff frequency. Bandpass filters pass only a limited range of frequencies and are specified in terms of a lower and an upper −3 dB frequency. Band-reject filters, also called notch filters, are just the opposite: They filter out frequencies within a given range and allow all others to pass.

---

[2] Hang on! We will get to the concept of the decibel shortly.

There are varieties of filters. We can make filters with steep or gradual cutoff slopes. We can make filters than affect only the phase of the spectrum—that is, the temporal alignment of the sinusoidal components—without affecting the overall shape of the spectrum. Analog filters are created using discrete components like resistors, capacitors, and inductors, and digital filters are created to run either in software on a computer or are embedded on a programmable digital signal processor (DSP) chip. This book is concerned only with filters that run in software on a computer, for they serve as a major tool in the toolbox of the digital audio workstation. We return to our discussion of filters in Chapter 4.

We have been casually talking about sound as both a temporal process and as a spectral process. As such, we can examine sound pressure variations in both time and frequency. If we measure the overall air pressure level of sound as a function of time, then we speak of a time-domain analysis, because we are looking at how sound changes over time. If we instead measure the frequency components that make up a sound during a given interval of time, we speak of this as a frequency-domain analysis. These two kinds of analyses are intertwined; we will return to the time-frequency duality nature of sound soon.

Sound and Human Perception

At the nexus of psychology, physics, acoustics, physiology, audiology, neurology, and music lies the vast realm of psychoacoustics and music cognition. We briefly touch on some of the basics here, because a basic understanding of how we hear and process sound can greatly assist your creative mastery of the modern digital audio workstation.

The ear can be decomposed into four functional regions:

1. Pinnae, or outer folds
2. Timpanic membrane (the "eardrum")
3. Ossicles (the "hammer," "anvil," and "stirrup")
4. Cochlea, or inner ear

The pinnae are connected to the timpanic membrane through the ear canal. Both the pinnae and the ear canal act as filters and aid in our localization of sounds in three dimensions. Depending on where sound is coming from, the pinnae and ear canal can filter it differently (just like an equalizer), and our brain can parse this information to help us discern spatial information about the sound source. The pinnae and ear canal are collectively known as the *outer ear*.

The timpanic membrane, commonly called the "eardrum," is quite simply a transducer (like a microphone) that converts acoustical energy (compressions and rarefactions of air pressure) into mechanical energy. Because the eardrum lies in contact with a series of bones, called ossicles, this mechanical energy is amplified and transmitted to the opening of the basilar membrane, a part of the cochlea. The tympanic membrane and ossicles are collectively known as the *middle ear*. The outer and middle ears taken together amplify incoming sound waves both mechanically and hydraulically by about 100 times.

The cochlea, or *inner ear*, is a spiraled, fluid-filled chamber. Because it is spiraled, like a snail, and filled with fluid, we can maintain our balance by perceiving where gravity is pulling the fluid. The basilar membrane lines the cochlea and is itself populated with many thousands of hair cells connected to auditory nerves. These nerve fibers are tuned like a filter bank to frequencies that we can hear (usually given to be 20–20,000 Hz, although the 20,000-Hz upper limit is a bit optimistic for most listeners) and transmit a signal when their corresponding frequencies are detected by the brain.

It is often said that the ear is a logarithmic instrument, meaning that it tends to respond to exponential differences in quantities (like amplitude and frequency) rather than arithmetic differences.[3] To this end, we have developed various ways of addressing our perception of sound intensity and frequency. One common way in which sound intensity is measured is with the decibel (dB).

The decibel is simply defined as

$$dB = 20 \times \log R \qquad (6)$$

where R is a ratio of two numbers. So to find out how many decibels bigger one signal is than another, we simply compute the ratio of their amplitudes, take the logarithm, and multiply by 20. Using your calculator, you can quickly see that ratio of 2:1 is then about 6.02 dB. A ratio of 10:1 corresponds to 20 dB. Table 1.2 lists commonly encountered decibel values and their corresponding ratios.

---

[3] There are various physiological regions for this, and I refer the interested reader to the "For Further Study" section at the end of this chapter.

R	dB Value
0.1	−20 dB
0.5	−6 dB
1	0 dB
1.5	3.5 dB
2	6 dB
10	20 dB
100	40 dB
1,000	60 dB
10,000	80 dB
100,000	100 dB

In audio circles, the *sound pressure level* (SPL) of a signal is measured relative to the threshold of human hearing. This threshold is simply the denominator in the ratio R from our last equation. Because sound is simply pressure waves in the air, one way to measure their intensity (as with a microphone) is to compute the amount of force (in Newtons, foot-pounds, dynes, etc.,) that the pressure wave exerts on a given area of space. The threshold of hearing is approximately 0.000002 Newtons per square meter. Thus, if we had a really, really big microphone whose diaphragm were 1 square meter in area, and we hit the diaphragm with 0.000002 Newtons of air pressure (that's not very much!), that would be the least amount of acoustical pressure we could possibly hear.

Scaling this by the actual area of the human eardrum (about 55 square millimeters), the least amount of air pressure our eardrums can pick up is about 1.1×10^{-10} Newtons, which as you could guess is not very much. That is roughly a millionth of a millionth pound per square foot! Anyway, this level is defined to be 0 dBSPL, and every other possible SPL is defined and computed relative to this level. Table 1.3 provides examples of typically dBSPL measurements for various activities.

The logarithmic perceptual nature of the ear pops up again in the ways in which we perceive frequency differences as well. As a reflection of this, ordered, periodic sets of notes, or *scales*, tend to be organized in geometric series in virtually all musical cultures. For instance, most scales throughout the world include an octave, or 2:1 frequency ratio. Thus, an octave above a 200-Hz fundamental is 400 Hz, an octave above that is 800 Hz, and so on. We tend to perceive octaves as exhibiting a special kind of pitch-class similarity, largely because frequencies an octave apart perfectly "line up" every second period, as shown in Figure 1.8.

TABLE 1.3

Example Sound
Pressure Levels
(dBSPL)

Sound Source	Approximate dBSPL
Threshold of hearing	0
Leaves rustling	10
Quiet whisper	20
Music studio	30
Normal conversation	50
Hotel lobby	60
Noisy office room	70
Machine shop	90
Train	100
Siren	110
Jet engine	120
Threshold of pain	130
Standing next to jet engine	140
Instant eardrum perforation	160–190
Space shuttle	215
Nuclear explosion	240–260

Figure 1.8
Octaves—that is,
2:1 frequency
ratios—occur in
the vast majority
of the world's
scales and tuning
systems.

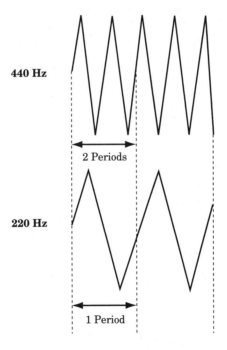

440 Hz

2 Periods

220 Hz

1 Period

Octaves form a geometric series built upon a factor of 2: Each successive member of the series can be obtained by multiplying the previous member by 2. Many scales in music are built as geometric series as well. One of the most common scales in use today, called *twelve-tone equal temperament*, forms a geometric series built upon a factor of the twelfth root of 2 ($\sqrt[12]{2}$). This has a particularly nice property for twelve-note scales: $\sqrt[12]{2}$, when multiplied by itself twelve times, yields 2 by definition, which is the octave ratio of which we just spoke. To obtain each successive member of the series, we simply multiply the frequency of the previous member by $\sqrt[12]{2}$, as shown in Table 1.4 for a twelve-tone equal-tempered scale built on A = 440 Hz. (By the way, $\sqrt[12]{2}$ is approximately 1.059463.)

TABLE 1.4

Twelve-Tone
Equal
Temperament

Pitch Class	Frequency Calculation	Resulting Frequency (Hz)
A	$440 \times 2^{0/12}$	440
A-sharp/B-flat	$440 \times 2^{1/12}$	466
B	$440 \times 2^{2/12}$	494
C	$440 \times 2^{3/12}$	523
C-sharp/D-flat	$440 \times 2^{4/12}$	554
D	$440 \times 2^{5/12}$	587
D-sharp/E-flat	$440 \times 2^{6/12}$	622
E	$440 \times 2^{7/12}$	659
F	$440 \times 2^{8/12}$	698
F-sharp/G-flat	$440 \times 2^{9/12}$	740
G	$440 \times 2^{10/12}$	784
G-sharp/A-flat	$440 \times 2^{11/12}$	831
A	$440 \times 2^{12/12} = 440 \times 2$	880

An infinite number of scales exist from which to choose. Even varieties of just intonation, which uses pure-tuned harmonic ratios to form scales, exhibits properties of the geometric series of octaves. But tuning and scales are a discussion for another day. I mention them here to illustrate the central role that logarithms play in music and perception. Awareness of the logarithmic nature of our perception of sound intensity and pitch can come into play particularly when mixing sound, to which we now briefly turn.

Mixing Sound

What does it mean to mix sound? We often say that mixing entails two different but interlinked activities: sound *transformation* and sound *layering*. By transforming either a recording of a physical sound (such as a guitar strum, a vocal line, or ocean waves) or a synthetic ("synthesized") sound, we can articulate particular features of the sound that we enjoy or want people to notice. We can also "repair" poor recordings to some extent, for example, by removing background noise or adjusting the pitch of a singer's voice. In transforming sound, we either do so for the sake of the sound itself or so that it will fit in the musical context of a particular mix. We will learn more about sound transformation in Chapter 5 when we discuss *plug-ins* for digital audio workstations.

Sound layering involves playing two or more synthetic or recorded sounds simultaneously. Layering can create certain problems if not done carefully. For example, layering sounds that contain energy in the same frequency band may result in a "muddy" mix in which the individual sounds are no longer identifiable.

But the real magic of mixing occurs when the sum is greater than the individual parts, and in this way, the idea of mixing is analogous to a conductor's leading a symphony orchestra. When we listen to an orchestra, we can choose to listen to individual instruments and their frequency content (what music cognitionists refer to as analytical listening). On the other hand, we can instead take in the entire experience as a combination not of individual instruments but rather new timbres that result from carefully orchestrated instrumental combinations. A well-crafted mix creates a new sound world from the carefully chosen and mixed individual components.

Studio mixing is a non–real-time process, as opposed to a live concert in which the audience hears sound immediately as it is produced. The studio has become a musical instrument, a melting pot for the recording, transformation, assembly, and sculpting of sound.

Studio mixing can serve several pragmatic purposes as well. For example, one voice can be recorded multiple times to create a chorus of voices. Musicians from around the world who have never met can play together. And a slightly out-of-tune singer can be made to sing in tune. Algorithms exist to do things like these in real time, but working out of real time in the studio uniquely affords fine-tuning of the details of these kinds of transformational opportunities.

Analog Warmth, Digital Precision

"Digital" has become something of a buzzword. We hear phrases like "digital quality" and "digitally recorded" so frequently and applied to so many different things that they seem meaningless. Particularly in the music world, analog and digital seem to have turned into philosophies or outlooks on life moreso that anything science or reason could explain. And to be fair, each way of thinking has its own benefits. But first, let's look more closely at each idea.

We use the term *signal* to describe some sequence or series of information-bearing quantities. Signals can be represented by a continuous line on a graph, a series of numbers, or a pattern of dots on a plasma screen. They can represent anything conceivable: musical pitches, the earth's magnetic field strength[4] seen by a particular satellite, or a student's grade point average as a function of time. And signals need not be functions of time; they can be functions of any dimension(s). However, as music is a function of time, most of the signals we are interested in here are functions of time (Figure 1.9).

Signals can be expressed in a continuous or discrete fashion. When signals are represented as continuous data, we describe them as analog. More formally, when each data point of a signal can assume an infinite number of values, and between any two data points (no matter how close) another value exists, we have an analog representation. In contrast, when signals are represented as discrete points of data, we describe them as digital. More formally, when signals can only assume a finite number of values, and when there exist gaps between values, we have a digital signal.

An analog representation of data is illustrated by a rotary knob called a *potentiometer*. We can turn the knob to any value we wish, and we can move it continuously. There are no gaps or jumps between values. A *rotary encoder*, or knob that "clicks" when turned, illustrates a digital representation. It snaps into place as we turn it; it can only assume certain predefined fixed values, and there are gaps between these values.

But a digital representation imposes one more constraint: Not only are a certain set of predefined values possible, but they are only allowed to occur at fixed points in time. This is why digital signals are also referred to as discrete-time signals. With digital signals, we quantize both the possible values a signal may assume as well as time itself. While these may seem restrictive, doing so affords an enormous power over the manipulation, analysis, and transmission of the signal.

Consider the advantages that each type of knob offers. The analog potentiometer allows continuous movement without jumps between

---

[4] See the "For Further Study" section for a musical take on this signal.

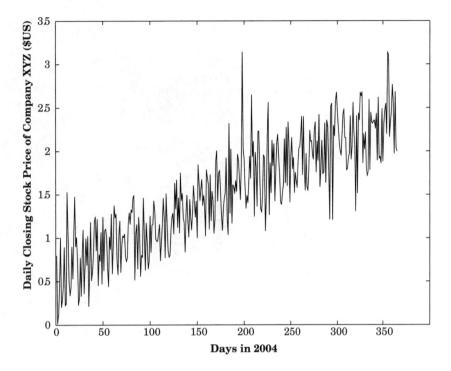

Figure 1.9
Examples of signals.

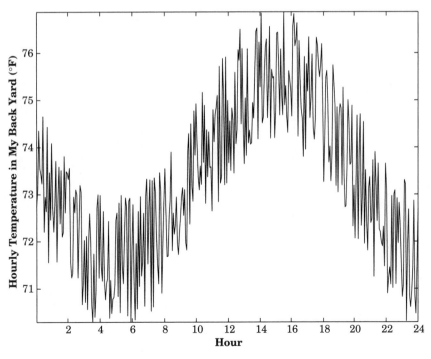

Figure 1.9
Examples of
signals
(continued).

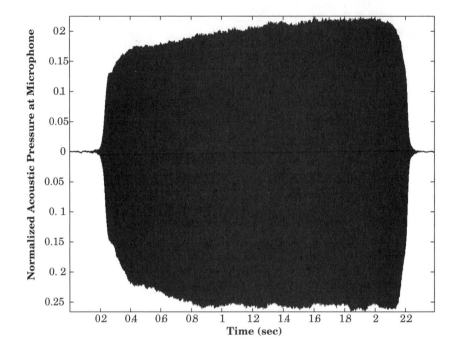

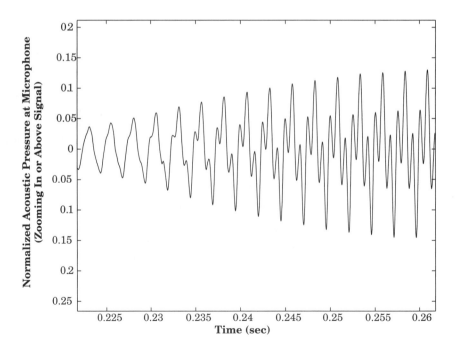

values. However, it is difficult to move it to an exact, repeatable location with a high degree of precision. Furthermore, potentiometers can be greatly affected by electromechanical noise when their internal contacts scrape against one another. The rotary encoder, by contrast, activates a corresponding one of its output pins when it is turned to a particular position. As a result, we can say with absolute certainty to which position the user has moved knob. The potentiometer's knob position can be more ambiguous.

Similarly, analog and digital representations of audio data each possess their own advantages and disadvantages. (See Table 1.5.)

TABLE 1.5

Digital Audio Representations versus Analog Audio Representations

Digital Audio Representations	Analog Audio Representations
Finite range of frequencies represented	Unlimited frequency range
Perfect transfers and copies	Degraded transfers and copies
Perfectly transmitted over longer distances (e.g., satellite radio)	Difficult to transmit over long distances without severe signal loss (e.g., FM radio)
Longer lifespan of storage media	Shorter lifespan of storage media
Unbounded theoretical dynamic range	Limited practical dynamic range
Manipulated inside computer	Manipulated usually by cutting and splicing tape
Lower noise floor easy to achieve	Higher noise floor
Nondestructive editing possible	Only destructive editing possible

As we will see shortly, digital audio is only capable of representing a finite band of frequencies, as opposed to analog audio, which can theoretically represent an unbounded frequency range. (The actual usable frequency range of analog audio is limited in the real world by storage media and transmission channels, however.) The frequency range of digital audio is theoretically unbounded, although the amount of data required quickly gets very large as we demand higher bandwidth. However, this finite frequency range need not be a disadvantage for digital audio: We can easily represent the range of human hearing and then some.

It's been said that the very high frequencies analog audio can represent may contribute to the so-called "warmth" of analog sound that seems missing from digital recordings. Consider, for example, the scraping of the horse hair of a violin bow against the strings of the instrument, which creates transient signals upwards of 50 kHz. Some argue that, although we cannot hear frequencies that high, they create audible frequencies through a process known as *heterodyning*. That is, we can likely hear the difference frequencies of adjacent high frequencies, so for example 50-kHz and 51-kHz components in the transient burst might be audible to us as a 1 kHz burst. Others contend that these effects are important only if you are a canine.

But the perceived "warmth" of analog recordings is more likely caused by differences in the ways that vacuum tubes and transistors operate. Analog recording equipment often employs vacuum tubes during the entire signal path, whereas digital recording devices generally employ only solid-state devices like transistors and diodes.

Although Thomas Alva Edison and John Ambrose Fleming had been working on technology that would ultimately lead to the vacuum tube, history generally credits Lee De Forest (1873–1961) with its invention as US Patent #979,275. (A related patent, filed in 1907, is US Patent #879,532.) De Forest, who filed his patent in 1905, called his invention an "Oscillation Responsive Device" and later the Audion, because he envisioned its use as an amplification device in audio circuitry. (See Figure 1.10.) The Audion was the immediate precursor to the vacuum tube triode and enabled early radio broadcast and sound amplification devices (Figure 1.11).

The vacuum tube, however, had its own problems. Chief among them was the amount of heat it produced, as it was essentially a light bulb with some extra goodies thrown inside, which led to a short life span for each tube. As Bell Labs engineer John Pierce (himself the father of modern satellite communication) noted, "Nature abhors the vacuum tube." In fact, Pierce coined the term "transistor" for a new device invented some 40 years after De Forest's patent by the team of physicists John Bardeen (1908–1987), Walter Brattain (1902–1987), and William Shockley (1910–1989) at Bell Labs as US Patents #2,502,488 and #2,524,035 (Figure 1.12). More or less functionally equivalent to the vacuum tube triode, the transistor began to replace tubes altogether owing to its much smaller size, longer life span, and lower production cost. Bardeen, Brattain, and Shockley were awarded the Nobel Prize in Physics in 1956 for their contributions.

After a bit of a hiatus, vacuum tubes seem to be in vogue once again. Actually, they never left the scene completely in audio circles, and many attribute this to their intangible "warmth" of sound. Many have investigated the differences between tubes and transistors in audio circuits, and the consensus seems to be that, when they are operated in their linear regions,[5] they are indistinguishable to the ear.

---

[5] You can think of this in two ways. First, the devices are operating in their linear regions when their outputs simply consist of their inputs multiplied by the constant scalar. Second, they are not turned up to eleven, as Spïnal Tap would say.

UNITED STATES PATENT OFFICE.

LEE DE FOREST, OF NEW YORK, N. Y., ASSIGNOR, BY MESNE ASSIGNMENTS, TO DE FOREST RADIO TELEPHONE CO., A CORPORATION OF NEW YORK.

OSCILLATION-RESPONSIVE DEVICE.

979,275. Specification of Letters Patent. **Patented Dec. 20, 1910.**

Application filed February 2, 1905. Serial No. 243,913.

To all whom it may concern:

Be it known that I, LEE DE FOREST, a citizen of the United States of America, and a resident of the borough of Manhattan, city, county, and State of New York, have invented certain new and useful Improvements in Oscillation-Responsive Devices, the principles of which are disclosed in the following specification and accompanying drawings, which explain the form of the invention which I now consider to be the best of the various forms in which the principles of the invention may be embodied.

My invention relates to an improvement in the sensitive member used in systems of wireless telegraphy to detect the electrical waves or oscillations and comprises the novel features hereinafter shown and described and particularly pointed out in the claims.

In the accompanying drawings I have shown, and in the description thereof will point out, certain forms of construction which may be employed in carrying out my invention and in connection therewith I will point out the principle of my invention.

Although by no means all the known or possible embodiments of my invention are herein illustrated or described, sufficient are given to make clear the principle of my invention.

Figures 1 to 6 inclusive each represents a receiving set for a wireless telegraph system, each having a sensitive member differing in appearance but all embodying the principle of my invention.

I have discovered that if two bodies adapted for use as electrodes or conductive members, be electrically separated partially or wholly, after the manner common in analogous devices, the separation between them may be neutralized sufficiently to enable them to act as a detector of electrical oscillations, if the intervening or surrounding gaseous medium be put into a condition of molecular activity, such for instance as would be caused by heating it in any manner, as by radiation, conduction, or by the combustion of gases in the space which surrounds the poles. Such condition or molecular activity causes what would otherwise be a non-sensitive device to become sensitive to the reception of electrical influences. I am thus enabled to employ as such sensitive member, devices which would otherwise be of no value, or to make those devices now used more sensitive to the electrical waves. This principle is embodied in the apparatus illustrated in the various figures shown.

In each of these figures, A represents the antenna or receiving conductor, or wave intercepting means; E, the earth connection; F, F¹, F², F³, F⁴, and F⁵, the electrodes used in their various forms; B the local battery; and T, the receiving or indicating instrument, which is herein shown as a telephone receiver, the same being shown only as typical of any form of indicating apparatus capable of being employed for such purposes.

In Fig. 1 the two electrodes F, F, are slightly separated and are within the flame of an ordinary Bunsen burner D. Under these conditions the electrodes may be adjusted so that there is normally no indication of a passing current given by the receiving instrument, such as the telephone T. The electrical separation of the electrodes is however insufficient to prevent electrical oscillations from jumping the gap. The influence of these oscillations upon the heated gas seems to break down or lower the insulating quality of the gap, so that, while the influence of the oscillations lasts, the current of the local circuit may pass between the electrodes, thus affecting the indicating instrument therein to produce a signal. This may be due to ionization of the gases surrounding the electrodes which greatly increases their conductivity, said ionization being more or less accomplished or greatly facilitated by their previous heating which has already put them into a condition of intense molecular activity.

In Fig. 2 the electrodes F¹, F¹, are of sufficiently great resistance to be heated by a current from a dynamo G, and by their radiation heat the gas between them. This gas may be air or the electrodes may be inclosed and surrounded by any suitable gas as shown, for example, in Fig. 6 in which said electrodes are inclosed in the receptacle H. The heating of the gas may also be by radiation from the electrodes F¹, F¹.

In Fig. 3, the electrodes assume the form of two parallel plates F², F², which are heated by a Bunsen burner. Although I have shown a Bunsen burner for this purpose, this particular heating device is not essential, as any means of heating may be employed.

In Fig. 4 the burner itself is made one

Figure 1.11
An audion from
ca. 1912. Photo
courtesy Mike
Schultz.

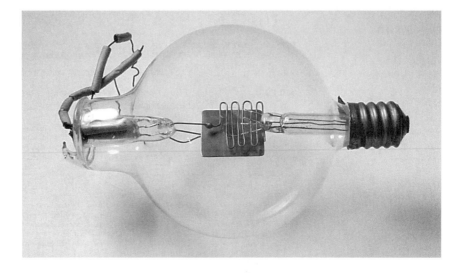

Figure 1.12
The first transistor,
created at Bell
Labs.

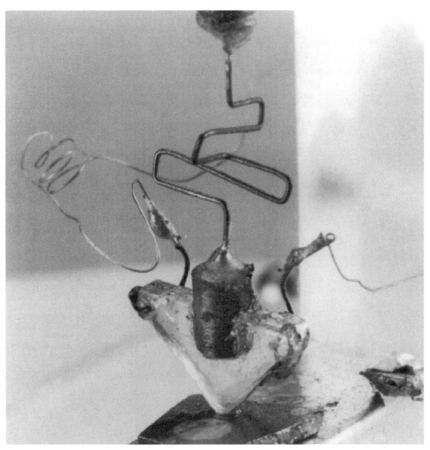

However, as they are driven into their nonlinear regions (i.e., driven to distort the input), measurable differences begin to appear. First, tubes exhibit what is commonly called "soft clipping": They very gradually begin to distort as they are driven "hotter," that is with increased input amplitudes. By contrast, transistors tend to exhibit "hard clipping," in which the input is quickly clipped. (See Figure 1.13.)

Figure 1.14 illustrates simplified *transfer functions* for vacuum tubes and transistors. The transfer function is an important engineering term that helps us characterize systems, whether they be implemented as physical hardware or in software. In loose terms, the transfer function of a system—a hardware effects processor, a software plug-in, a guitar stomp box—is simply the ratio of the output of the system to the input to the system. Transfer functions tell us what the system does to any input.

Figure 1.13
(a) Original signal;

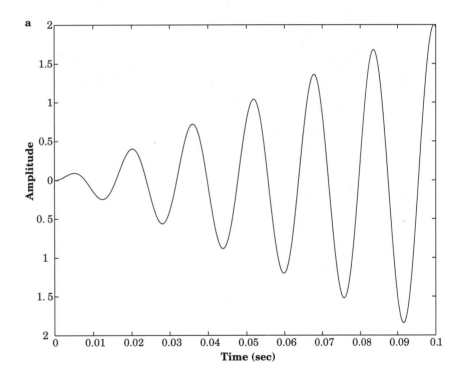

Figure 1.13
(b) soft-clipping
the signal; (c)
hard-clipping the
signal. Note the
more rounded
edges in the right
half of (c).

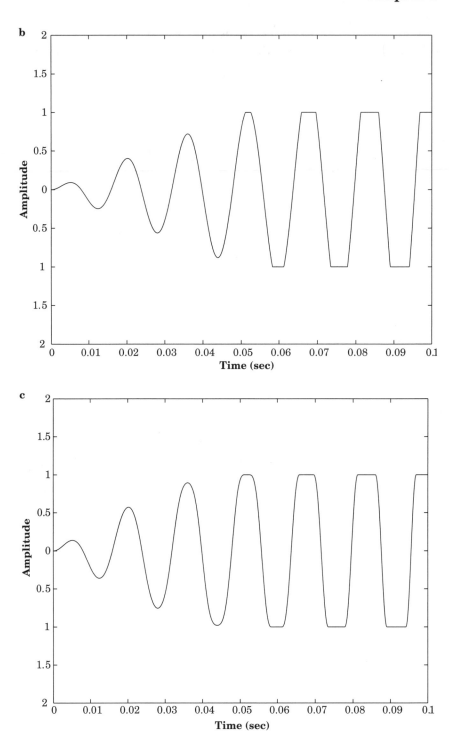

Figure 1.14
Typical transfer
functions of (a)
vacuum tubes
and (b) transistors.

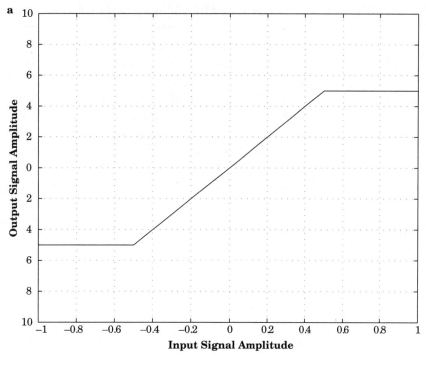

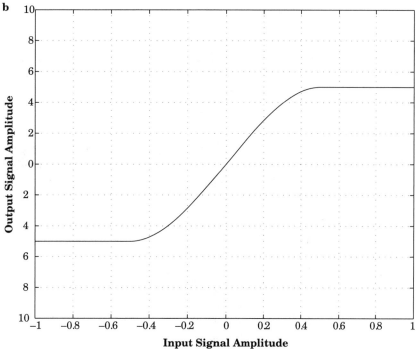

The horizontal axis in each graph represents the signal amplitude across the device's input terminals, and the vertical axis in each graph represents the corresponding output level. The graphs themselves illustrate each device's corresponding transfer function; that is, ratio of output to input.

The linear region of each device occurs where the transfer function is essentially a straight line. Whenever the transfer function deviates from linear operation, we have *nonlinear distortion* of the input signal. Distortion introduces harmonics to the output signal, and for this reason, amplifiers are often driven to distort or "thicken" the sound and make it timbrally richer. But tubes and transistors introduce harmonics in a very different way owing to their transfer functions. Audio engineers have discovered that transistor distortion can result in a very strong third harmonic component, which gives the resulting sound something of a harsh, biting quality. Tubes, on the other hand, tend to produce a richer set of harmonics simultaneously, particularly the second, third, fourth, and fifth. This tends to give the resulting sound a brass-like quality.

Another reason for the difference in sound between tubes and transistors involves the noise they introduce. Circuits based on vacuum tubes generally have a much higher noise floor (we discuss this term more later) than their transistor-based counterparts. Some say that this gives recordings made with transistor-based systems a more "edgy" or "cleaner" quality. Without that higher noise floor, every little detail of a mix can be present with no discernible noise to cover it up, and this may be a little more in one's face, as they say, than some desire. Indeed, this "edgy" quality led to much early skepticism and some hesitance in the adoption of the digital audio compact disc over analog LPs.

Again, modern studios have found clever ways to imbue digital studios with some of the benefits of analog sound, as it were. A popular way is to record all signals through a vacuum tube-based microphone preamplifier before converting them to digital form. Some record entirely to analog tape before mixing on a digital audio workstation. A friend of mine even manipulates all sounds with analog tape before mixing them inside her computer. Some modern-day mix engineers will only record onto analog tape, even though they may mix entirely inside a digital audio workstation. This effectively turns the analog recording chain into a kind of effects processor, because the analog process in not prized so much for its fidelity and storage or editing capability as for the inherent "feel{ it imparts to the sound.

Despite these differences, both analog and digital signals are typically represented by and transmitted with the electrical quantities of current or voltage. When communicating analog signals, we usually establish a range of values, say 0 to 5 volts, that our signal may take. (Again, it may take any value in that range.) That way, all members of the communication chain are aware of the lowest and highest possible values. This is the way most analog synthesizers communicated before digital communications were possible or commonplace; these voltage signals were called *control voltages*.

Digital audio workstations work exclusively with digital audio signals—sound waves whose pressure levels have been measured, recorded, and quantified for storage in a computer. When communicating digital signals, recall that we quantize both the values of the data and time itself. As we begin discussing this quantization of data and time into discrete values, and particularly as we delve into storing these quantized values inside a computer, we very quickly discover that we need to know a thing or two (or three) about binary numbers.

Taking a Byte Out of Bits

Analog systems store continuous data on an analog storage medium, like magnetic tape or a vinyl record. These media can store analog signals because it is relatively straightforward to modulate some physical aspect of the media (magnetism or groove position) in a continuous fashion. Analog recording and storage is loved by many owing to its unique sonic character; many also enjoy the tactile aspects of handling "sound" in their hands by manipulating it with razors, cutting and splicing it however they wish (Figure 1.15).

Digital systems, on the other hand, store data in series of *bits*, or binary digits, which can take the value of either 0 or 1. We call this numeric representational system base-2 logic, or simply base 2. (Some people have even proposed *trits*, or trinary digits, which can assume one of three values: 0, 1, or 2.) Given this, it is possible to store digital data virtually anywhere, because our storage medium need only be able to assume one of two values that we can change over time. We store digital data as a stream of ones and zeroes; we can send ones and zeroes using Morse Code over a telegraph, light pulses from a flashlight, or alternating high and low pits on the surface of a DVD.

Figure 1.15
The tools of analog recording and storage. (a) Otari two-track recorder; (b) Studer 24-track recorder;

Figure 1.15
The tools of analog recording and storage. (c) splicing block for cutting tape; (d) this fancy-schmancy splicing block makes editing and crossfades much easier!

In fact, numbers can be represented with any numeric base. It's most convenient for us carbon-based entities to use the base-10 (decimal) system, because we have been accustomed to doing so for centuries. Computers, however, usually benefit most from binary number systems, because they are built upon transistors that can act as extremely fast switching devices. Transistors can be used as either signal amplifiers (like a home stereo amplifier) or as signal switches and gates. When a transistor is used in an electronic circuit in a certain way, it can be made to switch "high" or "low" to represent 1 or 0, respectively, by turning it on or off.

But using a single bit to represent numbers won't get us very far. Clearly, using only single bits, we can only count from 0 to 1. To count any higher, we must group bits into chunks, just like we do in our decimal system. Consider the base-10 sequence {0, 1, 2, 3, 4, 5, 6, 7, 8, 9}. After 9, we must begin grouping numbers together to form larger numbers, so this sequence is followed by {10, 11, 12, ...} and so on.

The same thing happens with numbers represented in any arbitrary base. In binary logic, the sequence {0, 1} is followed by {10, 11} and then {100, 101, 110, 111}. This is shown graphically and extended in Figure 1.16.

Binary   **Decimal**

0000   0
    ↑
0001   1
↑
0010 0011   2   3
↑
0100 0101 0110 0111   4   5   6   7
↑
1000 1001 1010 1011 1100 1101 1110 1111 8   9   10 11 12 13 14 15 16
↑

Figure 1.16   *Counting in a binary numbering system.*

Note that we can arbitrarily pad as many zeroes before a number as desired. (For example, 23 in decimal is understood to have the same value as 0023.) So, in the example above, we count in each group by starting as far right as possible and running sequentially through all possible ordered numeric combinations in base 2 (0 and 1, as indicated by boldface). Next, we shift our attention left one position, turn what was formerly a 0 now into a 1, and run through all possible subsequent combinations of zeroes and ones to the right. You can also think of this process as flipping the value of the next-left bit in question and then appending the previous sequence, so {0, 1} is followed by {10, 11}. We repeat this again, and we've just counted from 0 to 15 in binary. Notice that, just as with decimal, the numbers get progressively longer as we count higher.

Bits can be combined together into *words* to form larger numbers. Computers usually group bits together in chunks of 4 called *nibbles* and chunks of 8 called *bytes*. Most computers further group bits into 16-bit words, some into 32-bit words, and still others into 64-bit words. We call the number of bits that make up a word the *word length*.

It is simple to convert numbers from any numeric base to any other numeric base. Let's introduce a notational convenience: To avoid confusion, we often indicate the numeric base of any number with a subscript after the number, so 1011 in binary could be written 1011_2. Now, consider how we intuitively compute the value of numbers displayed in decimal notation, for example, the number 333_{10}. The rightmost digit is in the one's position, so it is multiplied by 1_{10} to yield itself, 3_{10}. The middle digit is in the ten's position, so it is multiplied by 10_{10} to yield 30_{10}. And the leftmost digit is in the hundred's position, so it is multiplied by 100_{10} to yield 300_{10}. The results are added together to obtain $300_{10} + 33_{10} + 3_{10} = 333_{10}$. Suppressing the subscripts, we could rewrite this as

$$3 \times 10^0 + 3 \times 10^1 + 3 \times 10^2 + = 300_{10} + 33_{10} + 3_{10} = 333_{10}$$

Note that the number in each position as we move left is multiplied by a higher power of ten, starting in the zeroth (rightmost) position. The same is true for numbers of any base. Let's convert the binary number 1011_2 to decimal. Starting with the rightmost digit (bit) and suppressing the subscripts, we write

$$1 \times 2^0 + 1 \times 2^1 + 0 \times 2^2 + 1 \times 2^3 = 1 + 2 + 0 + 8 = 11_{10}$$

Sound into Numbers: Analog-to-Digital Conversion

Let's say we had a magical black box that could take analog signals at its input and spit out digital signals that corresponded in some way to its analog input. Millions of these black boxes already exist, and they're called analog-to-digital converters (or A/D converters for short). You can buy a simple one very inexpensively, but this was not always the case. Just a few decades ago, they were big, bulky, and expensive.

But A/D converters do not magically produce digital signals from analog input; they must be told how often to take a snapshot of, or *sample*, the continuous input, and they must be told how many bits to use in encoding the sample value. We call the frequency of sampling the *sample rate* (or *sampling rate*) and the number of bits used the *sample word length* (or just *word length* or *bit depth*). How might the sample rate and the sample word length impact the fidelity of the conversion?

Clearly, the more bits we use and the more often we sample the input, the more faithful the digital signal will be to the analog signal. This leads to two definitions. The difference between the quantized digital signal and the original analog signal is called *quantization error*. The ratio between the largest amplitude and the smallest amplitude we can encode using a given number of bits is called the *dynamic range*.

Quantization error results in distortion of the analog waveform we're trying to encode, because the analog signal must be "rounded" to the nearest digital quantization level at each sampling instant. Thus, if we are trying to digitize a perfect sine wave for example, it will end up looking a bit "stairstepped." The more quantization levels (i.e., bits) we use to encode the digital signal, the less "stairstepped" the digital signal appears. However, no matter how many bits we use, there will always be quantization error. Figure 1.17 illustrates the effect graphically of using larger word lengths when quantizing an analog signal.

A simple analogy might help our understanding of quantization error in digital-to-analog conversion. If we order a couple of pizzas that cost at total of $23.51 but we only have $20 bills, we either underpay the delivery person by $3.51 or overpay by $16.49—that's a potential error of 70%. If we have $10 bills, the most we could under- or overpay is $6.49, or 28%. The fact is, the smaller the denominations in our wallet, the smaller the error in our payment to the delivery person.

Fortunately, we can use some nifty engineering tricks to reconstruct the smooth, continuous analog waveform from the digitized, "stairstepped" one. We'll get to that in a minute.

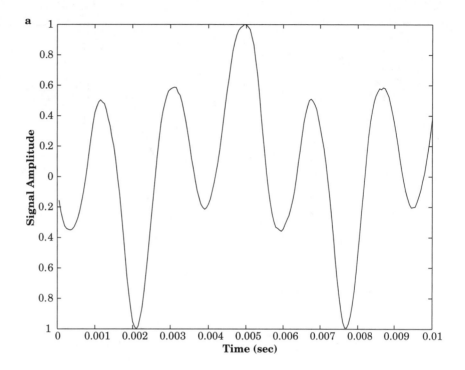

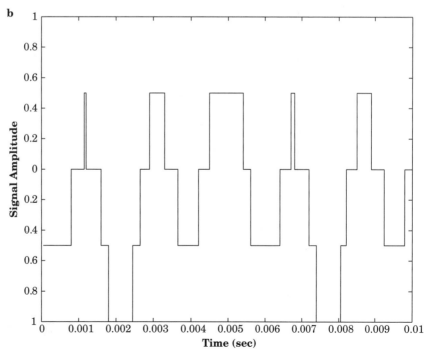

Figure 1.17
(c) 3-bit word
length digital
signal; (d) 4-bit
word length
digital signal.

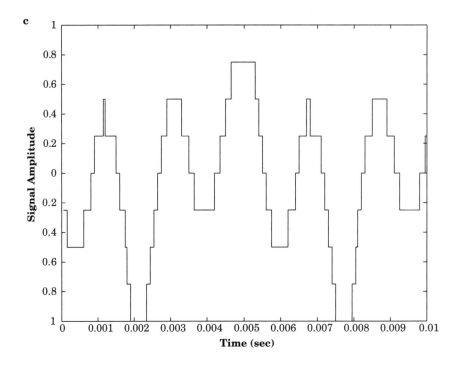

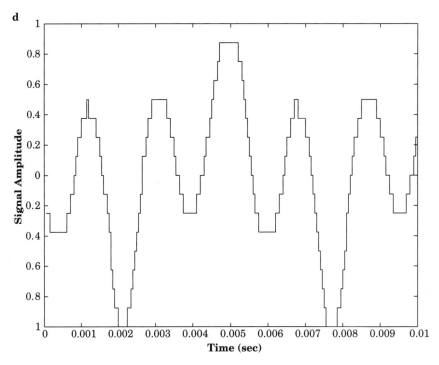

Let's zoom in on a single sample value to look at things more closely. Because our A/D converter simply rounds each sampled analog value to the nearest quantization level, it could introduce as much as a full bit of error. That is, analog signals that fall between ± one-half of a quantization level get rounded to that quantization level. (See Figure 1.18.) The effects of this rounding process introduces error— quantization error—into the digital audio signal.

Getting back to pizza, if we only had nickels, we could pay the $23.51 with an error of only a penny, or 0.042%. But only carrying around nickels comes with a price: carrying a bunch of nickels takes up a lot more space than only carrying a few $20 bills. Similarly, quantizing audio in 24 bits takes up a lot more memory in computers than does quantizing it in 8 bits, for example.

This leads right into another important term, the *signal-to-noise ratio*, which is a simple concept. Consider a symphony orchestra, which can produce around 110 dBSPL on stage in a large concert hall, assuming you're standing on stage with them. (Out in the audience would be softer.) And let's say the quietest level they might produce, called the *noise floor*, given an occasional rustling of music, coughs, and such, could be about 20 dBSPL. That's a difference of $110 - 20 = 90$ dBSPL. Thus, this particular orchestra exhibits a *signal-to-noise* ratio of 90 dB. (You might ask why we found this ratio by subtracting rather than by dividing. Recall that decibels are a logarithmic function, and dividing numbers turns into subtracting logarithms.)

To distinguish from the analog world, we often use the analogous term *signal-to-error ratio* when discussing the dynamic range of digital signals. Remember that each bit that we add multiplies the total available number of quantization levels by a factor of 2, which is $20 \times \log 2 \cong 6.02$ dB. Sweeping a few technical complications under the rug, this means we achieve about an extra 6.02 dB of dynamic range for each bit we use. The ratio of the maximum amplitude we can encode digitally to the quantization error level—that is, the signal-to-error ratio—is usually given in decibels by the equation

$$\mathrm{SE}_{\mathrm{dB}} \cong 6.02n + 1.76 \qquad\qquad (7)$$

where n denotes the word length. Thus, the 16-bit words that a standard audio compact disc uses yield around 98 dB of dynamic range, which is enough to capture the dynamic range of our symphony orchestra pretty well.

Figure 1.18
The difference between the original, analog signal and the quantized value is called quantization error. (a) The big picture; (b) zooming in;

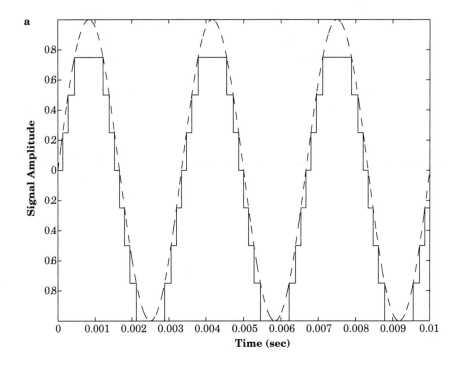

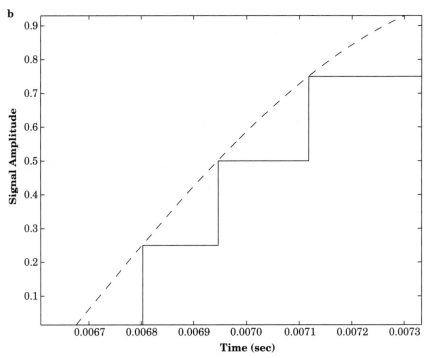

Figure 1.18
The difference between the original, analog signal and the quantized value is called quantization error. (c) corresponding quantization error (the arithmetic difference between the analog and quantized signals for 3-bit quantization; (d) corresponding quantization error for 4-bit quantization.

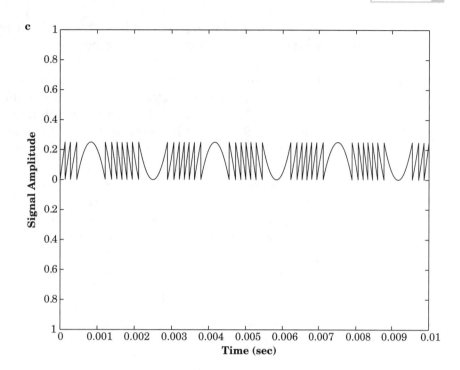

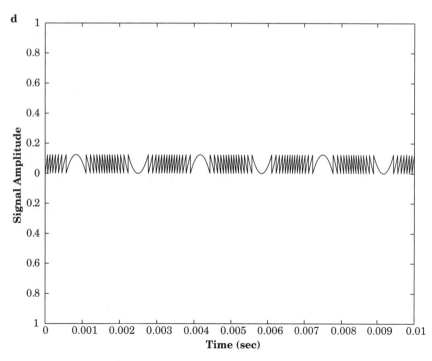

But wouldn't using extremely high sample rates and very large sample word lengths give us much greater accuracy? Why not just use extremely high sampling rates and word lengths? Well, we do in some fields like communications, but with audio signals we quickly reach a point of diminishing returns where our ears simply cannot tell the difference. Plus, a 10-GHz, 64-bit A/D converter is much more expensive than a 96-kHz, 24-bit converter! There are also some nifty tricks we can perform to make a relatively low-sample-rate converter and low–word-length converter sound great.

Specifically, what do increased sample rates and word lengths do for us? We'll first examine sample rates. According to the Nyquist Theorem, the sample rate of an A/D converter must be set to at least twice the highest frequency present in the analog signal to faithfully quantize the signal. So, the higher the sample rate of the converter, the higher the frequency the digital signal can represent. Said another way, the higher the sample rate of the A/D converter, the greater the *bandwidth* the digital signal can represent, because it can reproduce a broader range of frequencies. So, a 44.1-kHz compact disc can theoretically include a 22.05-kHz sinusoid, and a 96-kHz DVD-audio disc can contain a 48-kHz sinusoid.

Consider one cycle of a sinusoid, shown in Figure 1.19a. A minimum of two digital samples is required to represent each cycle. If we use fewer, an artifact known as *aliasing* results, when we attempt to convert the digital signal back to analog form. Aliasing occurs whenever one frequency in the actual signal is mistaken for another; in the example shown in Figure 1.19c, the frequency of the sinusoid is aliased, or confounded with, a lower-frequency sinusoid upon reconstruction.

Let's say we decide to use a sufficient sample rate to capture the sinusoid, and let's now consider the effect of the sample word length. The greater the number of bits we use to quantize the analog signal, the more discrete levels our digital signal can assume. Recall that a binary number with n bits can represent 2^n values. For example, a 2-bit binary number can assume 2^2 or 4 values:

$$00_2 = 0_{10}$$
$$01_2 = 1_{10}$$
$$10_2 = 2_{10}$$
$$11_2 = 3_{10}$$

Figure 1.19
(a) One period of
a sine wave. (b)
Sampling at least
at the Nyquist rate
is required.

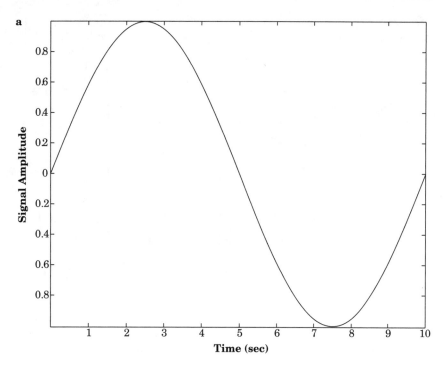

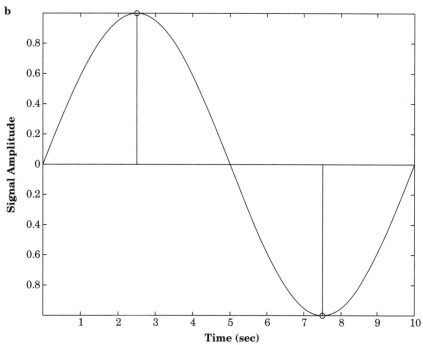

Figure 1.19
(c) Sampling
below the Nyquist
rate leads to
aliasing on
reconstruction.

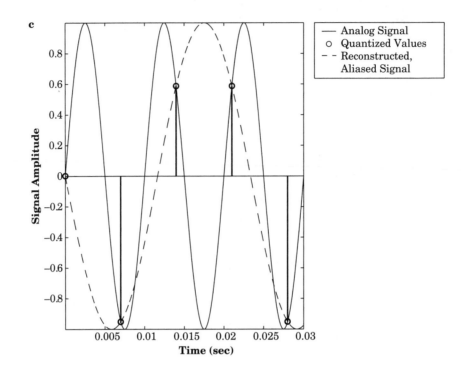

Table 1.7 explores the relationship between word length, quantization levels, and approximate signal-to-error (S/E) ratio for a simple digital audio encoding scheme.

TABLE 1.7

Word Length,
Quantization
Level, and S/E
Ratios

Word Length	Quantization Levels	S/E Ratio (dB)
1	2	7.8
2	4	13.8
4	16	25.8
8	256	49.9
12	4096	74.0
16	65536	98.0
18	262144	110.1
20	1048576	122.1
24	16777216	146.2
32	4294967296	194.4
64	18446744073709551616	387.0

Numbers into Sound: Digital-to-Analog Conversion

Once an analog signal has been transferred to the digital domain, a great wealth of processing functions becomes available. The signal can be transmitted, stored, duplicated, and processed with bit-for-bit accuracy. But once all the desired transmission, storage, duplication, or processing of the signal is completed, it needs to be converted back to analog form so that we can hear it. A digital-to-analog converter (D/A) is used for this purpose.

By the way, it is becoming increasingly common to wait as long as possible in the signal chain before converting digital audio back to analog form. Usually, the digital audio workstation (DAW) itself is used for both A/D and D/A conversion. However, the DAW is increasingly used to transmit digital audio without converting it back to analog; loudspeakers and even amplifiers are now available that accept digital audio input and produce analog output.

In the early days of the digital audio workstation, the D/A converter might be quite far removed physically from the A/D converter. In the Columbia-Princeton Electronic Music Center (1970s), which was a consortium between both universities, Princeton housed an A/D converter and computer, and Columbia housed the D/A converter. To hear the results of one's computer processing of musical signals, it was necessary to drive to Columbia and wait for the D/A to convert digital signals back to a form in which they could be heard. It certainly took a great deal of patience (and time) to make music with computers!

Fortunately, the situation is much different now. Digital-to-analog converters, like their modern A/D counterparts, are available on tiny integrated circuit chips. They are fast, inexpensive, and readily available. In fact, every modern computer contains both D/A and A/D converters, but their quality is generally not sufficient for studio-quality work with sound.

The Digital Audio Workstation

Because general-purpose computers do not include sufficient quality A/D converters for musical work, we typically use external audio hardware for this purpose. With a few modifications, computers can quite easily assume the role of a fully-equipped traditional production facility. A *digital audio workstation* (DAW) is a general-purpose computer that has been augmented with at least one separate audio interface and audio processing and mixing software. Digital audio workstations can cost anywhere from US$500 for a low-end general-purpose editing machine with minimal storage space to well over US$250,000 for a multitrack professional system with dedicated, outboard signal-processing hardware and multiple control surfaces.

Aside from the addition of one or more digital audio interfaces, hardware controllers, and accessories like microphones and loudspeakers, the modern DAW does not require any external audio-processing equipment. Using the power of a general-purpose computer to perform all signal processing, mixing, routing, mastering, and preparation of a disc master for reproduction can eliminate the need for purchasing the separate, special-purpose (and usually quite expensive) equipment typically found in traditional recording studios. This is a simple observation, and one with which not everyone may agree, but it has profound implications for the music industry on all levels, from composition to production, mixing to distribution. Consider, for example, the many recording labels that now purchase DAWs for their signees rather than renting studio time for them in commercial facilities.

A New Studio Paradigm

The notion of basing an entire music production studio around a central computer lies in stark contrast to the traditional paradigm of a studio. Simply put, the DAW can effectively replace and encapsulate much or all of the functionality present in a traditional console- and outboard-gear–based studio. The tradeoff is a fundamental paradigm shift of operability—a completely new set of human–machine interactions that one must learn to use the DAW, even though software for the DAW is firmly rooted in the traditional archetypes of interaction found in console-centered studios. A broad spectrum of studio technologies—both analog and digital—is available for all price points (Figure 1.20).

In general, DAWs can be further categorized by the kinds of audio interfaces attached to them. On the one hand are audio interfaces whose only function is to provide the computer with high-quality D/A and A/D converters. On the other hand are audio interfaces and

Computers, Sound, and Digital Audio

47

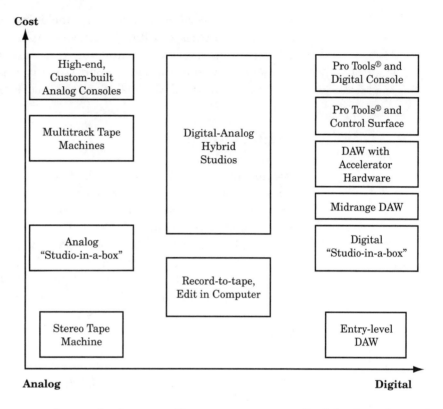

Figure 1.20
A broad spectrum of audio production tools is available.

expansion cards that actually assume some or all of the digital signal processing and number-crunching required for editing, mixing, and processing sound. These systems free the host computer to concentrate instead on running its operating system and managing files and disk access. Not surprisingly, these latter systems typically cost much more and can be more expensive to upgrade. However, the schism in processing power and capabilities is slowly decreasing as general-purpose microprocessors approach the signal-processing power of dedicated DSP chips. The gap is also being bridged by some mixing software that can take advantage of a signal-processing accelerator card if it is present but does not require one to run.

DAW Components

As mentioned, a general-purpose computer supplemented with outboard A/D and D/A converters lies at the center of the modern DAW. The DAW also includes additional hard disk storage, peripherals such as CD- and DVD-writers, and software including two-track sound file editors and multitrack editor-sequencers. Additional hardware like synthesizers and keyboards, and software such as audio plug-ins, sound synthesis environments, and video-processors and editors, often supplement the DAW, making it even more robust.

Purchasing a DAW

Musicians have several options available when deciding to establish a digital audio workstation facility, whether in a project studio at home or in a professional music studio. One option is to simply purchase a computer and appropriate software, attach an audio interface, and integrate all the elements together. A second option is found in cus-tom-made, tested, and integrated "turnkey" systems available from third-party vendors. The latter type of pre-built DAW offers conven-ient, easy setup and generally includes technical support for the entire DAW system; the trade-off is they cost more than building a DAW from scratch.

Disadvantages of the DAW

The digital audio workstation presents certain advantages and disad-vantages relative to the traditional outboard-gearbased studio, which accounts for much of the friendly debate between users of DAWs and traditional audio production equipment. And of course, a disadvan-tage to one person might be an advantage to another.

The first disadvantage of the DAW cited by some is the simple fact that it operates completely within a computer—that one is tied to using a computer to make music. That is, the audio production is at the "mercy" of the computer's operating system and hardware. This potentially requires frequent updating of both software and hardware components. And unfortunately, no operating system is completely crash-proof. By contrast, analog mixing consoles (and most digital ones for that matter) just don't crash!

Another disadvantage, at least until recently, has been the lack of human interface options for the DAW. Musicians accustomed to mix-ing on analog consoles and working with knobs, buttons, and switches yearn for immediate, tactile control of the mix, and a computer key-board is not quite the same.

Advantages of the DAW

Many different interfaces for working with DAWs have recently become available. From miniature "knob boxes" and jog shuttles to full-scale 96-fader control surfaces, the interfaces with which we can control DAWs quite effectively augment the computer keyboard and improve our work efficiency.

The fact that a general-purpose computer lies at the heart of the DAW is also the very source of its strength. Software versions of functionally identical hardware boxes such as reverberators, delay units, and equalizers often cost much less than their physical counterparts. And the wealth of sound-processing and synthesis tools currently available, from freeware and shareware to high-end professional software, provides an unbounded workspace—an infinitely open musical canvas.

And even though traditional mixing consoles and outboard processing gear do not "crash" like a computer can, their upkeep and repair can be far more laborious, expensive, and time-consuming, and they can wear down over time. The typical maintenance of a DAW involves occasional software and operating system upgrades for the host computer. As a result, the total cost of ownership of DAW-based studios is typically far less than their console-and-gear studio counterparts.

And Moore's Law, which predicts that the number of transistors on a silcon chip doubles every couple of years, ensures the continued survival and advancement of the DAW as the heart of an increasing number of both project and professional studios. In terms of value, quality, total cost of ownership, maintenance, and expandability, a well-designed and configured DAW simply cannot be beat.

For Further Study

Aristotle. *De Anima* [On the Soul], tr. J. A. Smith.

Chain Reaction (1996). Twentieth Century Fox. Directed by Andrew Davis.

de Poli G, Piccialli A, Roads C, eds. *Representations of Musical Signals*. Cambridge, MA: MIT Press; 1991.

Dodge C. "Earth's Mangetic Field." Nonesuch LP H-71250.

Hamm RO. "Tubes versus Transistors—Is There an Audible Difference?" *Journal of the Audio Engineering Society* 1973;21(4):267–273.

Hong S. *Wireless: From Marconi's Black-Box to the Audion*. Cambridge, Massachusetts: MIT Press; 2001. http://www.leedeforest.org.

Moore G. "Cramming More Components onto Integrated Circuits." Electronics 1965;38(8).

Negroponte N. *Being Digital*. New York: alfred A. Knopf; 1995.

Pohlmann KC. *Principles of Digital Audio*. New York: McGraw-Hill; 2000.

Putterman SJ. "Sonoluminescence: Sound into Light." *Scientific American*, February 1995.

Rumsey F. *The Audio Workstation Handbook*. New York: Butterworth-Heinemann; 1996.

This Is Spinal Tap (1984). MGM/UA Studios. Directed by Rob Reiner.

http://www.lucent.com/minds/transistor/history.html.

Exercises and Classroom Discussion

1. An ideal tuning fork tuned to A = 440 Hz, when moderately struck, only vibrates at 440 Hz. In the section "Sound and Human Perception," we casually mentioned that it would tend to vibrate at integer multiples of 440 Hz when struck harder. Why is this true? What other things vibrate at non-integer multiples of a given frequency when struck or hit?

2. Audio distortion ("fuzz" or "grunge") circuits can be made from vacuum tubes or diodes. What features of each explain the sonic differences of the result?

3. When might a mix benefit from "hard clipping" of a track? What about "soft clipping"? Explain the how these two types of distortion sound different from one another.

4. Is it possible to store "CD-quality" digital audio on a vinyl LP record? If so, how would you do it?

5. Listen to the composition *I Am Sitting in a Room* by composer Alvin Lucier, recorded and mixed using analog tape, which exploits the acoustical characteristics of a physical space to construct an abstract sound world. Reconstruct this idea digitally by recording a physical sound source into your computer; play it back out a loudspeaker, and record the result again into your computer. Repeat this process until the original source is deconstructed and unrecognizable.

6. Apply *dither* to a copy of a sound file while viewing the original file in another window on your computer. Zoom in on individual samples in each file and comment on their visual and sonic differences.

7. What kind of A/D converter would you rather have in your studio: a 64-bit 44.1-kHz converter, a 16-bit 192-kHz converter, or a 1-bit 10-Mhz converter? What are the advantages of each? Why?

8. (a) Count to 7 in base 5. (b) What is $3_3 + 5_4 + 7_2$? (c) Is it possible to represent a number in base π?

9. What is oversampling, and why is it used in your CD player?

10. One sound was recorded at 44.1 kHz, and another sound was recorded at 192 kHz. What must you do before mixing them? How would you do it?

11. Investigate and categorize the six physical phenomena that result in the production of sound waves.

12. What audio interfaces are currently available for laptop computers? What is the maximum number of input/output channels available? What is the maximum bit depth and sample rate available?

13. What is the maximum number of audio tracks that one can record in real time onto a desktop computer? Include specific references to software and hardware.

14. Pick out a pop tune that you think exemplifies a good mix and another that exemplifies a bad mix. Explain what contributes to your assessment of each.

15. What built-in functions and programs does your computer's current operating system include for working with sound files?

16. Listen to some "glitch" music (for instance, by the Japanese sound artist Ryoji Ikeda) or perhaps a work by Radiohead or Aphex Twin. How would you describe the work to someone else who has not heard it? Is it possible to create such a music using strictly analog means?

17. Where does the number 1.76 come from in Equation 7?

18. Assemble an approximate budget for two studios—one based on a traditional console, and one based around a DAW—that have approximately the same mixing and processing capabilities.

The Development of the Modern DAW

Before we delve into the features and capabilities of the modern digital audio workstation, it is important to understand the precursors and technologies that have enabled its very existence. Understanding something about the past—and especially the hurdles that had to be overcome—is crucial to a successful investigation of future potential.

As with any history of computing, attempts to dissect hardware developments from software developments are difficult, if not impossible. Developments in each area lead to advancements in the other; they are inexorably intertwined. However, it does make some sense here in this chapter to consider dedicated hardware-based precursors to the modern digital audio workstation separately from entirely software-based systems designed for general-purpose computers. A large gray area exists between the two, but this chapter in its entirety should provide the big picture.

The technical abilities of the modern DAW are the direct result of research and development in academic computer music centers, federally funded speech/telephony research projects, commercial music studios, and commercial audio hardware/software companies. Each camp was of course interested in different things: academia and commercial studios demanded the greatest audio fidelity, speech and telephony desired the least number of bits to transmit and hence the highest data compression rates, musicians demanded ease of use, and companies needed to be profitable. Thrown into this quandary of desires was the research, development, and compositional use of sound synthesis algorithms for computers, something to which each of the above groups contributed significantly.

During the early days of computing, the primary method used to manipulate sounds was cutting and splicing magnetic tape. A experimental music movement known as *Musique Concrète*, which began in Paris the late 1940s, with roots in the Italian Futurist movement and the German Dadaist movement, led to the creation of many techniques and methods for tape manipulation that would later become more or less commonplace in recording studios around the world. *Musique Concrète*, a movement spearheaded by Pierre Henry and Pierre Schaeffer, and later growing to include many others, developed alongside the earliest digital audio developments. These techniques and their associated terminology—cutting, looping, reversing, flanging, scrubbing, and so on—have been transplanted to digital audio workstations, and are still in use today. Instead of cutting and splicing tape, however, we just cut and splice binary numbers (or, more likely, use a software program that does the number-crunching for us!).

It is a very daunting task to provide a complete and thorough account of the development of the modern DAW, let alone in one chapter. I have attempted rather to highlight just a few of the major turning points in this wonderful history of making music with computers by letting those involved speak for themselves wherever possible. For brevity, some major developments and pioneers are not included; I hope this chapter serves as but a starting point for further explorations into the historical coexistence of music and technology. An excellent source for further historical study is Joel Chadabe's book *Electric Sound: The Past and Promise of Electronic Music.*

This necessarily brief history of software and hardware for making music with computers is continued in Chapter 9, when we discuss current extensions to the DAW, many of which are rooted in the traditions presented here. I hope that the historical foundations that laid the groundwork for the modern DAW presented here will not only impart a sense of the historical context of computers and music but will instill a deeper appreciation for the tireless work that has enabled the DAW to be what it is today.

It is especially important to note the importance of the community and environments in which these tools were created and continue to be created. As with virtually all musical instruments—and ultimately any musical endeavor that involves technology in some way—a fundamental collaboration between engineers, composers, and performers often makes the difference between historical success and failure.

Bell Labs

Although the CSIRAC computer (Figure 2.1) developed at the University of Melbourne was probably the first digital computer to produce sound from a set of instructions (as early as 1951), it did not include a D/A converter to translate digital pulses into analog waveforms. Rather, the computer had a small loudspeaker to which digital pulses were directly sent; the time distance between individual pulses controlled the resulting frequency.

In the 1950s, Max Mathews worked as Director of the Behavioral and Acoustic Research Center at AT&T Bell Telephone Laboratories in Murray Hill, New Jersey. Now frequently referred to as the "grandfather" of computer music, we might quite accurately here call him the grandfather of the digital audio workstation. In 1958, he

Figure 2.1
The CSIRAC
computer (1949).
Photo courtesy
Paul Doornbusch.

began writing computer software to synthesize audio, or generate sound from scratch.

In 1965, an album was released on the now-defunct Brunswick record label featuring music synthesized and mixed at Bell Labs on an IBM 7090 computer and an attached sound card, called the "digital to sound transducer," an early D/A converter. The liner notes to this album speak for themselves, and they are reproduced here in their entirety.

Music from Mathematics

Played by IBM 7090 Computer and Digital to Sound Transducer

The course of human development has always been marked by man's [sic] striving for new techniques and tools in pursuance of a better life. This is most dramatically manifested in the fields of science and technology. But this dissatisfaction with available materials and methods and the corresponding search for new ones is also evident in the arts, and artists have continually sought to improve the tools of their trade. Today's modern orchestral instruments, for example, hardly resemble their medieval ancestors. On this recording, we illustrate another advancement in the realm of tools available to the music-maker: the computer and the digital-to-sound transducer. This new "instrument" combination is not merely a gadget or a complicated bit of machinery capable of producing new sounds. It opens the door to the exploration and discovery of many new and unique sounds. However, its musical usefulness and validity go far beyond this. With the development of this equipment carried out at the Bell Telephone Laboratories, the composer will have the benefits of a notational system so precise that future generations will know exactly how the composer intended his [sic] music to sound. He will have at his command an "instrument" which is itself directly involved in the creative process. In the words of three of the composers whose works are heard on this recording:

Man's music has always been acoustically limited by the instruments on which he plays. These are mechanisms which have physical restrictions. We have made sound and music directly from numbers, surmounting conventional limitations of instruments. Thus, the musical universe is now circumscribed only by man's perceptions and creativity.

The process of composing music on and for the computer and transducer is highly complex: we shall attempt here only a brief and simplified description so that the listener may better understand what he is hearing. At the very heart of this type of composition rests this fundamental premise: "Any sound can be described mathematically by a sequence of numbers." Our composer thus begins by determining what numbers specify the particular sounds in which he is interested. These numbers are then punched on IBM cards: the cards are fed into the computer and the digits recorded in the memory of the machine. The computer is thus able to generate limitless sounds, depending on the instructions given it by the composer. The latter, instead of writing the score in notes, programs his music by punching a second set of IBM cards, which when fed into the computer cause it to register on tape certain sounds from its vast storehouse. The composer may give the computer detailed instructions for every "note," or he may allow it varying degrees of freedom by asking it to select "notes" at random from a host of possibilities. The tape which emerges from the computer contains music in the form of magnetic impressions. To convert these impressions to actual sounds, the tape is run on a digital-to-sound transducer, which translates the digital indications to sounds, amplifies these sounds, and gives us the finished musical product.

As can be gleaned from the above description, the human element plays a large role in computer music, as in any art medium. The sounds and sound-producing methods are new; the composer's role is essentially that which it has always been. History tells us that whenever a new concept emerges, it is labeled revolutionary by either its proponents or the public at large. The new techniques and tools of computer music are not meant to replace the more traditional means of composition and performance. Rather, they are designed to enhance and enlarge the range of possibilities available to the searching imagination of musicians. Science has provided the composer with new means to serve the same ends—artistic excellence and communication.

Remember that scene in the film *2001: A Space Odyssey* in which the HAL 9000 computer sings "Bicycle Built for Two"? That bit of audio was created at Bell Labs during this time: Max Mathews programmed the accompaniment on an IBM 7094 computer, and John Kelly and Carol Lockbaum programmed the singing voice.

Later work at Bell Labs included the "Music-N" family of software-synthesis environments, which we will talk more about later in this chapter. The GROOVE (Generated Real-time Operations On Voltage-controlled Equipment) system represented another major development. This system incorporated unique input devices (like joysticks, knobs, and so on) and a Honeywell DDP224 computer connected to an analog synthesizer.

In 1977, an engineer at Bell Labs, Hal Alles, built the Alles Synthesizer, an all-digital real-time synthesizer. The Alles Synthesizer was developed into a commercial version called the GDS Synthesizer, which later turned into the Synergy and Mulogix line of digital synthesizers.

The RCA Synthesizer

The RCA Mark II Synthesizer (Figure 2.2), although not itself based on a computer, laid some foundation for the later digital audio workstation. A programmable "complicated switching device to an enormous and complicated analog studio hooked into a tape machine," according to composer Milton Babbitt, it was installed at the Columbia-Princeton Electronic Music Center. Completed in 1955 and designed by RCA engineers Harry F. Olson and Herbert Belar with a price tag of US$10,000, the Mark II was perhaps the first programmable, integrated system that could synthesize and sequence sounds. In 1959, composers Vladimir Ussachevsky and Otto Luening at Columbia University and Milton Babbitt and Roger Sessions at Princeton University obtained a Rockefeller Grant of US$175,000 over five years to purchase the RCA Mark II and establish the Columbia-Princeton Electronic Music Center.

Figure 2.2
The RCA Mark II
Synthesizer. (a)
Schematic; (b)
data-entry
keyboard.

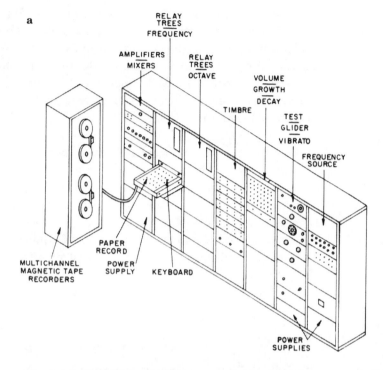

a

RELAY
TREES
FREQUENCY

AMPLIFIERS   RELAY
MIXERS   TREES
                OCTAVE

                        VOLUME
                        GROWTH
                        DECAY

                TIMBRE

                                TEST
                                GLIDER

                                VIBRATO

                                    FREQUENCY
                                    SOURCE

PAPER
RECORD

MULTICHANNEL   POWER
MAGNETIC TAPE   SUPPLY   KEYBOARD
RECORDERS

                                POWER
                                SUPPLIES

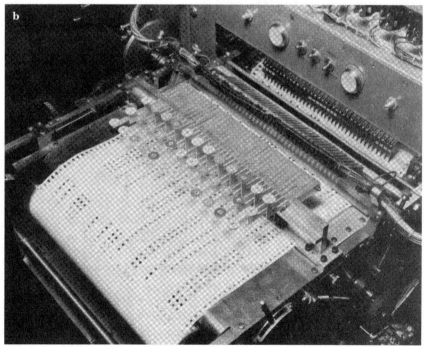

b

There was no computer keyboard on the Mark II Synthesizer; rather, it was programmed via a typewriter-like punch-card system, which can be seen in the bottom left of Figure 2.2. Figure 2.3 shows a sample punch card on the left and the musical information the card represents on the right. The punch card was read and deciphered by a mechanical system of brushes and switches that malfunctioned frequently. Milton Babbitt noted:

> The machine was extremely difficult to operate. First of all, it had a paper drive, and getting the paper through the machine and punching the holes was difficult. We were punching in binary. The machine was totally zero, nothing predetermined, and any number we punched could refer to any dimension of the machine. There was an immense number of analog oscillators, but the analog sound equipment was constantly causing problems.... I became irritated with the mechanics of the machine very often.... And yet for me it was so wonderful because I could specify something and hear it instantly.

Figure 2.3
RCA Punch Tape.

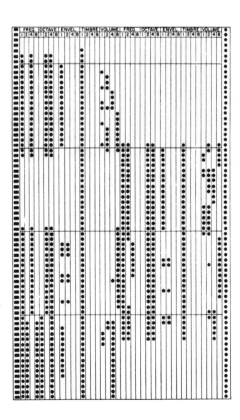

The punch-card system instructed the RCA Synthesizer to play specific sounds at specific times, and the results were recorded onto magnetic tape for later playback. The synthesizer contained sine-wave oscillators and noise generators, and these could be patched in different configurations to create different timbres. The RCA Synthesizer's tuning was limited to the twelve-tone equal-tempered scale. The Mark II was used at Columbia University until it was destroyed by vandals in 1976.

The University of Illinois

The University of Illinois at Urbana-Champign (UIUC) formed one of the most fertile breeding grounds for young digital audio workstation technology. In 1955, Lejaren Hiller (1924–1994) and Leonard Isaacson co-composed the *ILLIAC Suite for String Quartet* (his fourth string quartet) using the ILLIAC I computer (Figure 2.4), which is credited as the first composition for instruments composed algorithmically using a digital computer and often credited as the first true piece of "computer music."

Hiller, a chemistry professor at the time, earned a master's degree in music while teaching, and he founded the Experimental Music Studios at UIUC that same year, in 1958. His work on computers and music continued, leading to a programming language called MUSICOMP, written by Hiller and his colleague Robert A. Baker and based on many of the ideas behind information theory. In 1963, Hiller and Baker composed the *Computer Cantata*, a large-scale work in which the computer was used to algorithmically compose not only notes for instruments to play, but also texts for singers and electronic sounds as well.

Pioneering and creative audio work at the University of Illinois continued with the composers Salvatore Martinaro (1927–1995) and Herbert Brün (1918–2000). Martinaro and his colleagues developed a real-time performance instrument called the Sal-Mar Construction (see Figure 2.5), which unified a digitally controlled analog synthesizer, sequencer, composing machine, and 24-loudspeaker playback system. The instrument was controlled via 291 touch-sensitive switches that allowed the performer to zoom dynamically in and out of sequences, navigating quickly back and forth between the micro- and macro-levels. It weighed over 400 pounds fully assembled, and it

Figure 2.4
Lejaren Hiller at
the ILLIAC
Computer.

stood over eight feet tall. Martinaro wrote of his experiences with the Sal-Mar:

> The most oft-asked question after a concert performance with Sal-Mar was: "Do you know what it will do next?" Though too complex to analyze, it was possible to predict what sound would result, and this caused me to lightly touch or slam the switch as if this had an effect on the two-state logic. I was in the loop, trading swaps with the logic. Let's face it, there are some things you can't talk about and make much sense. I enabled paths, or better, I steered. To make music with the SMC is like driving a flying bus.

Composer Herbert Brün joined the faculty in 1962, and worked extensively with computers and music. His design for a sound-synthesis language called SAWDUST was realized by 1976, and it ran on a Digital Equipment Corporation PDP 11/50 computer. It allowed users to manually specify the time-domain shapes of waveforms. Brün wrote of this program:

Figure 2.5
Figure 2.5
The Sal-Mar
Construction.
Photograph
courtesy Dorothy
Martinaro.

The computer program which I called SAWDUST allows me to work with the smallest parts of waveforms, to link them and to mingle or merge them with one another. Once composed, the links and mixtures are treated, by repetition, as periods, or by various degrees of continuous change, as passing moments of orientation in a process of transformations.

SAWDUST has been updated and ported to other operating systems, and it is still runs on modern computers.

A number of other advances in modern DAW technology grew out of work at the University of Illinois as well. James Beauchamp, a professor with joint appointments in music and electrical engineering, developed a powerful and flexible software synthesis environment with Scot Aurenz called Music 4C, rooted in the "Music-N" tradition, that was used for many years. Two students, Carla Scaletti and Kurt Hebel, developed a digital audio system based on an early Apple Macintosh computer called Kyma. This soon led to a robust, real-time DAW that included a graphical programming language and the Capybara signal-processing hardware, known collectively as the Kyma System (1990), which is also hosted on a Macintosh computer.

The Legacy of Synthesizers

In the first decades of using computers to make music, the number-crunching requirements for audio were generally far beyond what the technology of the day could achieve in real time. As such, synthesizing and mixing sound was an extremely long and arduous process, often taking many hours or even days before one could hear and evaluate the audible result.

Digital synthesizers began to emerge, tracing their roots to the analog synthesizer as well as custom-made, electronic, real-time performance instruments. Analog synthesizers, such as those made by Robert Moog, grew out of a desire to literally perform music using electronic means. These synthesizers trace their roots to earlier electrical and electronic instruments from the earlier part of the twentieth century, like the Telharmonium (1897), the Optophonic Piano (1916), the Clavier à Lampes (1927), and the Hammond Organ[1] (1935), which tended to incorporate their music-making circuitry into a more or less traditional instrument form factor and played like a piano-style keyboard.

Real-time performance instruments also contributed to the birth of the digital synthesizer. Machines like Salvatore Martinaro's Sal-Mar Construction, Raymond Scott's Electronium (1950–1970), Marvin Minksy's Muse (1971), and the synthesizers made by Donald Buchla also grew out of a desire to perform (and now even compose) music in real time, rooted in the tradition of early electronic instruments not based on pre-existing form factors, like the Theremin (1920), Trautonium (1928), Dynaphone (1928), Ondes Martenot (1928), Rhythmicon (1930), Electronic Sackbut (1953), and many others.

---

[1] Incidentally, the Hammond Organ is credited as the first, large-scale, commercially successful electrical or electronic musical instrument.

An increasing number of computer-controlled analog synthesizers were developed. Using digital circuitry for control and analog circuitry for synthesis was quite appealing: Digital logic was by then well understood, as was analog sound-making circuitry. Using analog circuits to control analog synthesizers proved increasingly limiting to many composers; furthermore, digital logic more easily facilitated the notion of the sequencer. And using a computer to drive an analog device meant that control data could be stored and recalled. Examples of this type of device include Peter Zinovieff's Synthi 100 prototype (1968), which used a PDP-8 computer to control analog synthesizers. Max Mathew's GROOVE system represents another example.

Several new digital studios were founded in the early 1970s, while such work continued at Bell Labs, the University of Illinois, the University of Toronto, the Columbia-Princeton Electronic Music Center, and elsewhere. New synthesizers designed by Peter Zinovieff in England, Donald Buchla in the United States, and others incorporated digital technology extensively.

In the 1970s, two high-end digital keyboard-based synthesizers emerged that in many ways revolutionized the synthesizer as we now know it: the Synclavier and the Fairlight CMI. The Dartmouth Digital Synthesizer, the moniker of the Synclaiver prototype, was developed in 1975 and commercially available by 1976, from the newly formed company New England Digital (NED). The Synclavier represents the first synthesizer that used digital means to synthesize sounds. For only US$80,000 in the late 1970s, one could purchase a Synclavier with floppy disk drive, a 10-Mb hard disk, VT100 terminal display, an operating system, and some software. New England Digital later made significant contributions to the development of the modern DAW with the introduction of the widely used DAW called the NED PostPro Tapeless Studio hard disk recording system in 1979.

The Fairlight Computer Music Instrument (CMI) was released in 1979, by the Australian company Fairlight, its 1978 prototype being known as the QASAR M8. The CMI is credited as the first sampling instrument, allowing recordings of sounds ("samples") to be played from the keyboard rather than relying solely on synthesis techniques. Equipped with a monochrome video display unit, the user could even draw on the screen with a special lightpen waveforms to be played from the instrument's piano-style keyboard.

Winham Lab

An offshoot of the Columbia-Princeton Electronic Music Center came to be centered at Princeton, developing into the Winham Lab (see Figure 2.6), named after faculty music theorist and programmer Godfrey Winham. In 1964, Winham and Hubert Howe adapted the Music IV language from Bell Labs, a later version of Max Mathews' original Music I program, to an IBM 7094 computer housed within the university; the sound synthesis and composition language was written in assembly language and became known as Music 4B. Another version, Music4BF, was written in the scientific computing language FORTRAN.

Figure 2.6
Composer Paul Lansky at the Winham Lab (1981). Photograph courtesy Paul Lansky.

Composer Paul Lansky writes of the early days in the lab:

We used to carry digital tapes here from the IBM mainframe across the street. This room was basically used just to listen to the results of computations. ... Directly in back of me is the Hewlett-Packard 2116 computer we used for D/A conversion. It had 64k memory and no operating system (at least as we used it). The D/A program was loaded by paper tape. I believe the computer cost about US$20,000 when it was bought in 1973 or so. It was the first computer HP made, I think. On top of the computer you can make out a fan. This was used to cool it down. In the mid-1970s, the new television show NOVA filmed part of an episode about bird song in the lab, involving an interview with a researcher who was synthesizing bird songs on the computer. The room got terrifically hot because of all the lights the crew brought in, and the computer never recovered from the overload. (We frequently had to resort to opening the front of the computer—it was built like a refrigerator!) To the right of the computer is an 800BPI vacuum column tape drive. These huge tapes stored about 8 minutes of signal (probably the equivalent of about 4 minutes at 44k used today). (Out of the camera's view is a 1600 BPI drive. This stored twice as much sound!) To the left of the 2116 is a slightly newer HP computer. Mark Zuckerman and Godfrey Winham wrote a program called MOM (Music on Mini), which ran on this machine. To the left again is a Scully two-track tape machine, and in front of me are the smoothing filters, a variable pulse generator to set the sampling rate, a frequency counter, and a Heathkit amplifier Rick Cann built one summer. The lab closed down about two years later when we switched to using a PDP-11/34 for conversion. We sold the HP machines to a scrap dealer. It's hard to imagine, but everything in this room now fits into a portable CD player the size of your hand. The room was deafeningly loud when all the equipment was turned on. Lights flashed and tape drives whirred and the music went round and round.

Clearly, computers and software have come a long way since then, and thankfully so!

Minicomputer-Based DAWs

Frustrated with the computational limitations and real-time performance capabilities of computers, several pioneers in the mid-1970s created DAW systems based around minicomputers interfaced with custom signal-processing hardware. Two examples of such systems include the SYTER in France and the SSSP Synthesizer in Canada.

After the Groupe de Recherches Musicales (GRM) acquired a PDP-11/60 computer in the mid-1970s, work on the SYstème TEmps Réel/Realtime System (or the "SYTER," as it came to be called) began. Finished in 1982, The SYTER incorporated a home-made digital synthesizer that was controlled by the PDP-11/60 host computer, and it was commercially available by 1985. Able to manipulate, synthesize, and record to disk in real time, the SYTER and its functionality laid the groundwork for the popular modern DAW software plug-ins called the GRM Tools.

The SSSP Synthesizer was developed at the University of Toronto in the late 1970s. Although formed in 1959 as an analog electronic music studio, the studio obtained a PDP-11/45, and Bill Buxton and colleagues wrote software to combine the computer with custom-made sound-synthesis hardware and even prototypical graphical user interfaces.

Systems such as these laid the groundwork for DAWs today that are built around a host computer but that also rely on external signal-processing hardware to offer greater audio horsepower than the computer alone offers.

Home Microcomputers and the "Proto-DAW"

Although earlier computers intended for home use had been introduced, Apple Computer ultimately revolutionized the computer market with the introduction of the 1-MHz Apple I Computer kit in 1976 for around US$500. Not only was the computer affordable for many home users, it set the stage for what would be become the now-common personal computer form factor: a monitor and keyboard that sit together on a desk.

A small handful of home computers, notably the Yamaha CX5 Music Computer (1984) and the Atari 520ST (1985) included built-in MIDI ports for communicating with a synthesizer and rudimentary synthesis capabilities. But most early home computers did not include built-in sound cards, or if they did, they were extremely primitive. Several companies began to manufacture separate, standalone audio interfaces for early home computers like the Atari, Commodore Amiga, and the Apple II family.

A pioneer in this movement of bringing high-quality digital audio processing to the personal computer was a company called Micro Technology Unlimited, founded by Hal Chamberlin and David Cox. David Cox summarizes the early days of the confluence of personal computers and music:

> Dr. Robert Moog was the pioneer of the analog synthesizer. His vision-ary work is recognized by all who participated at the leading edge of music synthesis. However, these devices had to be adjusted hourly to keep them in tune. Creating a composition with them was tedious at best.
>
> The roots for the DAW started from the commercial needs for precise control of audio on computers; government funded speech research, commercial telephone research, and university computer music synthe-sis centers. Only the music centers needed 16-bit quality A/D–D/A con-verters running at CD quality sample rates. The first professional DAWs were mini and mainframe computers with audio A/D–D/A con-verters added for audio input/output. The status "father of digital audio" is granted to Max Matthews, who was at AT&T doing speech research for telephony.
>
> In the late '70s, university computer musicians wanted digital syn-thesis on their desktop, instead of dealing with university mainframe accounts. This created a small but exciting digital audio market. In 1977, MTU shipped the first digital music synthesis software and D/A converter boards for 6502 microcomputers (KIM-1, SYM-1, AIM-65, OSI, PET, Apple). All other companies with plug-in boards and software used square wave oscillators that created sounds like a "bee in a tin can." MTU was the technology leader, shipping wavetable synthesis products in 1977.
>
> A typical Apple II workstation would crash hourly with more than three expansion cards. This was unacceptable because some profession-al clients needed to compute uninterrupted for twelve to twenty hours to generate one high quality song. The software techniques MTU evolved allowed creating as complex a musical piece as desired, by drop-ping from real-time on the computation. Thus, a slow system (early microcomputers were slow) could, over tens of hours, compute a perfect orchestral.
>
> Thus, MTU learned in the '70s that absolute reliability was manda-tory to get the work done. Early microcomputers were inexpensive rela-

tive to minis and mainframes, but lacked the disk speed to record or process digital audio at professional sampling rates. Thus, in 1979, MTU developed and shipped the world's fastest floppy disk controller and software for audio playback—37.5KB/second sustained speeds. For example, Corvus Concepts Inc. hard drive for Apple II was only 8KB/second, 1/5th MTU's sustained speed, and MTU clients had inexpensive, removable media! Other comparisons are: Apple floppy disk—900 bytes/sec, IBM floppy disk—3Kbytes/sec, HP floppy disk—192 bytes/sec for program, 1.5Kbytes/sec for data. It is clear that MTU was the leader in fast disk transfers.

The first professional quality direct-to-disk digital audio workstation with 16-bit dynamics converters on a microcomputer was shown by MTU in 1979 at the West Coast Computer Faire in Los Angeles, CA.

Since the mid 1960s, MTU's founders understood the advantages of digital synthesis, the problems of obtaining the audio output (and input) hardware, and the requirements for shielding radio frequency interference (RFI) for the sensitive analog-to-digital (A/D) and digital-to-analog (D/A) converters. Thus, in 1982, MTU pioneered the DigiSound-16 externally shielded input/output (I/O) module for A/D and D/A converters. From 1982 to 1988, it was interfaced to DEC, SUN, MTU, IBM, and Apple Macintosh II computers, all used in professional audio work. Today, the accepted standard for professional products is to shield and remove the audio converters to an external box outside the computer.

Using the standard computer operating system and drive formats was very difficult. For example, on a SUN-3/160 computer, recording 16-bit stereo at 44.1 KHz required 90% of the computer's resources! Thus, most digital audio workstations even today use proprietary hard-disk formats to speed up transfers. In 1981 to 1986, MTU developed a series of microcomputer workstations. Using a proprietary file format, we could obtain incredible sustained transfer speeds from floppy disks. However, it was proprietary and required MTU to interface to every device that came along. Thus, it was not truly "open architecture"... allowing other products to be added easily. Today, using various standard operating systems such as Microsoft Windows, each peripheral device manufacturer ensures their hardware works with Windows. This creates a truly open architecture workstation...if the standard operating system drive and file formats are used.

In the late 1970s, we were delivering A/D/A converters for Kim-1, SYM-1, AIM-65, Apple, and PET 6502 microprocessor-based "DAW predecessors," and also to mainframe and super-minis used at universities worldwide doing music research and composition. In fact, the very first professional DAWs were used in government-funded speech research used for sonar and the CIA [the United States Central Intelligence Agency]. This was the advent of professional music workstations. They were cumbersome, but did the "digital audio" job. The hard drives used in those days, such as CalComp were "dishwasher size" for 300 MB. The second company to come on the market was IMS (Integrated Media Systems), whose owner/founder developed an A/D/A box for Stanford University called the Samson Box. He later sold to Studer and his descendent product was the Dyaxis DAW. We predated them in developing quality audio converters.

In the early 1980s, other companies developed synthesizers built around relatively inexpensive home computers. These include the Alpha Syntauri, which included a piano-style keyboard controller and audio interface card that relied on the Apple II for audio processing and synthesis, and the Rhodes Chroma (1992), also built around an Apple II. But the Apple was not a panacea, as engineer and musician Philip Dodds wrote in a 1982 article in an interview with Robert Moog in *Keyboard Magazine*:

> When it comes to recording and processing digital representations of continuously varying control voltages, the Apple itself is the limiting factor. What we need is "multitasking capability," the ability to transfer data from semiconductor memory (RAM) to disk while other data is pouring into or out of the computer, or is being processed. There are accessory cards for the Apple that can do that, but we do not have the necessary software yet. We are also looking to more powerful personal computers, like the new IBM or the DEC personal computers, that are capable of multitasking without the addition of extra accessory hardware.

Rather than designing hardware to retrofit existing home computers into digital audio workstations, some sought software-only solutions to sound-file mixing, editing, and manipulation, even if real-time operation had to be forsaken. We will talk later in this chapter about historically significant software-based mixing programs, but one software-only suite not really used for mixing but worth mentioning here is the Composer's Desktop Project (CDP). Formed in the mid-1980s, and spearheaded by composer Trevor Wishart, the CDP, which exists to this day, represents a consortium of musicians and programmers who devote time to a unified, open-source distribution of audio-processing tools. Initially accessible only via a command-line interface, the CDP sound-processing library has grown to include several graphical user interfaces for Windows computers, and it includes an enormous suite of sound-editing and processing algorithms. The CDP package grew to incorporate a wealth of both time- and frequency-domain processing tools.

Still others yearned for a new generation of completely extensible, robust, and open frameworks for the real-time synthesis, manipulation, and mixing of digital audio, even at the expense of adding expensive outboard gear to existing computers. Two options seemed to be emerging: either wait for faster computers and microprocessors that could handle all the number-crunching, or build custom hardware based on one of the new digital signal processors that were becoming both available and inexpensive.

The Samson Box and Stanford University

The Center for Computer Research in Music and Acoustics (CCRMA) at Stanford University has been involved with the development of modern digital audio workstation since its founding in 1975. Known primarily for its development of software synthesis environments in the Bell Labs "Music-N" tradition, it is also famous for the original sound patent on frequency modulation (FM) synthesis, by Professor John Chowning, which was licensed to Yamaha and became the basis for their popular DX-7 digital synthesizer. In fact, the FM patent has remained one of the most lucrative patents ever issued to Stanford.

CCRMA commissioned a digital signal-processing hardware box from a company called Systems Concepts, which delivered the device in 1977. As Stanford professor Julius O. Smith writes:

> While the Music V software synthesis approach was somewhat general and powerful—a unit generator could do anything permitted by the underlying programming language—computational costs on a general-purpose computer were dauntingly high. It was common for composers to spend hundreds of seconds of computer time for each second of sound produced. Student composers were forced to work between 3 A.M. and 6 A.M. to finish their pieces. Pressure mounted to move the primitive sound-generating algorithms into special-purpose hardware.

In October 1977, CCRMA took delivery of the Systems Concepts Digital Synthesizer, affectionately known as the "Samson Box," named after its designer Peter Samson. The Samson Box resembled a green refrigerator in the machine room at the Stanford Artificial Intelligence Laboratory, and it cost on the order of US$100,000. In its hardware architecture, it provided 256 generators (waveform oscillators with several modes and controls, complete with amplitude and frequency envelope support), and 128 modifiers (each of which could be a second-order filter, random-number generator, or amplitude-modulator, among other functions). Up to 64 Kwords of delay memory with 32 access ports could be used to construct large wavetables and delay lines. A modifier could be combined with a delay port to construct a high-order comb filter or Schröder all-pass filter—fundamental building blocks of digital reverberators. Finally, four digital-to-analog converters came with the box to supply four-channel sound output. These analog lines were fed to a 16-by-32 audio switch that routed sound to various listening stations around the lab.

The Samson Box was an elegant implementation of nearly all known, desirable, unit-generators in hardware form, and sound synthesis was sped up by three orders of magnitude in many cases. Additive, subtractive, and nonlinear FM synthesis and waveshaping were well supported. Much music was produced by many composers on the Samson Box over more than a decade. It was a clear success.

The Samson Box, however, was not a panacea. There were sizable costs in moving from a general software synthesis environment to a constrained, special-purpose hardware synthesizer. Tens of man-years of effort went into software support. A large instrument library was written to manage the patching of hardware unit generators into instruments. Instead of directly controlling the synthesizer, instrument procedures written in the SAIL programming language were executed to produce synthesizer commands that were saved in a "command stream" file. Debugging tools were developed for disassembling, editing, and reassembling the synthesizer command-stream data. Reading and manipulating the synthesizer command stream was difficult but unavoidable in serious debugging work. Software for managing the unique envelope hardware on the synthesizer was developed, requiring a lot of work. Filter support was complicated by the use of 20-bit fixed-point hardware with nonsaturating overflow and lack of rounding control. General wavetables were not supported in the oscillators. Overall, it simply took a lot of systems programming work to make everything work right.

Another type of cost was incurred in moving from the general-purpose computer to the Samson Box. Research into new synthesis techniques slowed to a trickle. While editing an Algol-like description of a Mus10 instrument was easy, reconfiguring a complicated patch of Samson Box modules was much more difficult, and a lot of expertise was required to design, develop, and debug new instruments on the Box. Many new techniques such as waveguide synthesis and the CHANT vocal synthesis method did not map easily onto the Samson Box architecture. Bowed strings based on a physical model could not be given a physically correct vibrato mechanism due to the way delay memory usage was constrained. Simple feedback FM did not work because phase rather than frequency feedback is required. Most memorably, the simple interpolating delay line, called *Zdelay* in Mus10, was incredibly difficult to implement on the Box, and an enormous amount of time was expended trying to do it. While the Samson Box was a paragon of design elegance and hardware excellence, it did not provide the proper foundation for future growth of synthesis technology. It was more of a music instrument than a research tool.

The Samson Box was integrated with several computers and used for quite some time at CCRMA. Synthesis languages like Garreth Loy's MBox (1979) and William Schottstaedt's Sambox (1979) ran on a PDP-10 computer augmented with the processing capabilities of the Samson Box until the late 1980s, when CCRMA abandoned the PDP-10.

Later work at CCRMA included a DAW developed by William Schottstaedt and Heinrich Taube that was based on a NeXT computer. The computer's signal-processing power was augmented with an Ariel Corporation Quint Processor containing five dedicated Motorola digital signal processors. Work there continues, most recently in developing DAW-tailored distributions of the Linux operating system for off-the-shelf computers.

The IRCAM Signal Processing Workstation

Giuseppe Di Giugno's groundbreaking digital synthesizers known as the 4A (1976), 4B (1977), 4C (1979), and the 4X (1981) also established a firm tradition of hybridizing general-purpose computers and custom-built audio hardware by harnessing their collective potential into an integrated system. The 4X Workstation, developed at the Institut de Recherche et Coordination Acoustique/Musique (IRCAM) in Paris, combined several computers and a 4X Synthesizer into one package that cost around US$100,000 at the time, and it represented one of the pinnacles in real-time digital performance instruments of its day.

Work began on a technological successor to the 4X Workstation based on a newer, more powerful general-purpose computer. A team at IRCAM chose the new NeXT computer as the host for the IRCAM Signal Processing Workstation (ISPW). The ISPW retrofitted a signal-processing card containing a Motorola 56001 DSP and two Intel i860 DSPs inside the NeXT computer, incorporating much of the 4X's functionality within a faster, much more powerful and extensible, musical workstation. (Figure 2.7).

Figure 2.7
The IRCAM Signal Processing Workstation (ISPW). Photograph courtesy Cort Lippe.

Other Workstation-Based Systems

The 1980s gave rise to a new generation of powerful computer workstations that were based on the UNIX operating system. To this extent, many others also began to leverage the power of these computers in creating customized digital audio workstations by combining them with custom A/D and D/A converters. One such system, based on the PCS/Cadmus Workstation computer, was introduced in 1984 for a cost of about US$35,000. Another system, known as the Interim DynaPiano, incorporated a Sun SPARCStation computer running UNIX, DSP cards, audio production software, and an audio interface into a single environment.

Next-Generation Commercial Systems of the 1980s

Parallel to the development of UNIX workstation-based audio workstations primarily in academia, several companies were formed to actively design and produce commercially available, integrated DAWs. Many of these systems were so well designed that they are still in widespread use today, almost two decades after their initial deployment.

Developed by a team of engineers led by James A. Moorer (a CCRMA graduate) in the early 1980s, the SoundDroid Workstation is widely acknowledged as the first digital, nonlinear hard-disk–based audio editor, although only one prototype was built. Moorer, working for Lucasfilm Droidworks, designed a custom digital-signal processor (called the ASP, or Audio Signal Processor) around which the workstation was based. It was used extensively in Lucasfilm movie productions, such as *Indiana Jones and the Temple of Doom* and *Return of the Jedi*. Complete with a trackball, touch-sensitive displays, moving faders, and a jog-shuttle wheel, the SoundDroid included programs for sound synthesis, digital reverberation, recording, editing, and mixing. The company eventually folded, and former employees founded a company called Sonic Solutions, which would go on to produce the first 24-track DAW and the first DVD-authoring environment.

Other digital audio workstations followed, including the AMS AudioFile and the Lexicon Opus, which exclusively used custom-designed hardware. Systems like the Digital Dyaxis (1988) piggy-backed on an Apple Macintosh computer. But another system in particular was about to ignite a veritable DAW revolution, the products of which dominate the modern DAW market.

A company called Digidesign was founded in 1983 to develop digital-audio editing tools. In 1985, they released SoundDesigner II editing software, arguably the standard two-channel computer audio editor for the next decade or more. Their 1989 Sound Tools system hybridized a Macintosh computer with custom signal-processing hardware and an audio interface into a US$3,995 package. In 1991, Digidesign unveiled its Pro Tools suite of hardware and software for digital audio editing, mixing, and processing, which again was designed for the Macintosh. A complete Pro Tools system offered similar functionality to the popular NED PostPro at about one-tenth the price.

The Legacy of Software-Based Mixing

The basis of the modern digital audio workstation's very existence is the idea that sounds—whether recorded or synthesized—can be edited and mixed entirely in the digital domain. This idea of entirely software-based mixing found roots in both academia and industry, and the legacies of several key pieces of software from each group have empowered the modern DAW in various ways. Once it became feasible to process and mix sound files on a computer—even if not in real time—many pioneers began writing their own audio software for general-purpose computers, leading to what we now think of as the digital audio workstation.

The history of the DAW is intimately intertwined with the history of digital sound synthesis, because the tools for synthesizing sounds with a computer developed into general-purpose sound-processing and sound-mixing environments. These environments paved the way for software-based mixing a fundamental aspect of the modern DAW. Many of these tools were released as open-source software (i.e., the source code is freely available) to the community.

The "Music-N" Family

Working at Bell Labs in New Jersey in 1957, Max Mathews wrote the first program (entirely in IBM 704 assembly language) for synthesizing and mixing sound on a computer, called Music I. This groundbreaking program evolved into an entire generation of music software: Music II (1958), Music III (1960), Music IV (1963), and eventually Music V (1969), which was written in FORTRAN on an IBM 360 computer. These computer music environments are collectively known as the "Music-N" family, and the code was developed and shaped by several others. We spoke earlier about Music4B and 4BF; other offshoot developments included John Chowning's Music 10 (1966), developed at Stanford, and Lejaren Hiller's MUSIC7, developed at SUNY Buffalo, also in the late 1960s.

Primarily known for their sound-synthesis capabilities, the later members of the Music-N family enabled users to mix sound by loading sound files into wavetables and playing the tables back simultaneously. The relative playback levels of the sound files could be controlled via breakpoint envelopes, a precursor to the breakpoint automation found on many consoles and DAW programs today.

A particularly important adaptation of the Music-N family was produced by Barry Vercoe at MIT. Rooted in his earlier programs Music 360 (1969) and Music 11 (1973), the Csound software environment was released in 1986 and is still very much in use today. In fact, Csound itself lies at the core of the modern MPEG-4 synthesis specification standard. Vercoe tells the history of Csound in Richard Boulanger's excellent tutorial, *The Csound Book*, referenced at the end of this chapter:

> This field has always benefited most from the spirit of sharing. It was Max Mathews' willingness to give copies of Music 4 to both Princeton and Stanford in the early '60s that got me started. At Princeton it had fallen into the fertile hands of Hubert Howe and the late Godfrey Winham, who as composers imbued it with controllable envelope onsets...while they also worked to have it consume less IBM 7094 time by writing large parts in a BEFAP assembler (Music4B). Looking on was Ken Steiglitz, an engineer who had recently discovered that analog feedback filters could be represented with digital samples. By the time I first saw Music4B code (1966–67) it had a reson filter—and the age of subtractive digital sound design was already underway....
>
> But we were still at an arm's length from our instrument. Punched cards and batch processing at a central campus facility were no way to interact with any device, and on my move to the Massachusetts Institute of Technology (MIT) in 1971, I set about designing the first comprehensive real-time digital sound synthesizer, to bring the best of Music 360's audio processing into the realm of live interactive performance. After two years and a design complete, its imminent construction was distracted by a gift from Digital Equipment Corporation of their latest creation, a PDP-11. Now, with a whole computer devoted exclusively to music, we could have both real-time processing and software flexibility, and Music 11 was the result....
>
> On returning to MIT in 1985, it was clear that microprocessors would eventually become the affordable machine power, that unportable assembler code would lose its usefulness, and that ANSI C would become the lingua franca. Since many parts of Music 11 and all

of my Synthetic Performer were already in C, I was able to expand the existing constructs into a working Csound during the fall of that year.

As with its predecessors, Csound is generally run by specifying an *orchestra*, which defines the particulars of each synthesized instrument that will be used, and a *score*, which tells the instruments in the orchestra when and how to play. Although initially a non–real-time environment, today's faster computers can easily run most Csound code instantly in real time, allowing the user to hear the results as they are computed.

A simple example of a Csound orchestra and score is given below.

```
/* diskin.orc */
; Initialize the global variables.
sr = 44100
kr = 44100
ksmps = 1
nchnls = 1
; Instrument 1: play an audio file
instr 1
  asig diskin "liam.wav", 1
  out asig
endin

/* diskin.sco */
; Play the audio file for five seconds.
i 1 0 5
; end
e
```

CARL

Newer post–Music-V software synthesis and mixing languages were developed in the early 1980s at the University of California San Diego's Computer Audio Research Laboratory (CARL). In 1982, CARL's director, F. Richard Moore, and D. Gareth Loy released a program called cmusic as part of an entire open-source distribution of UNIX-based music-making tools. Build on the C programming language and leveraging UNIX's powerful C-shell, the CARL Distribution has been developed, extended, and faithfully maintained over the years, forming the basis for many other powerful audio workstations.

MIX/Cmix and Rt

"Mix" is a name that has been used by a couple of different programs. A graphical mixing program known as Mix was written by Øyvind

Hammer of the Norsk nettverk for Akustikk, Teknologi og Musikk (NoTAM) in the 1990s. It is also the name of an earlier, text-only program for mixing sound written by Paul Lansky of Princeton University in the late 1970s.

"Cmix" is also a name shared by two different programs. The most recent is a free command-line mixing program for Linux. The first program to bear the name *Cmix*, however, was written by Lansky as a UNIX version of his MIX program. Arguably the most important contribution of Mix and Cmix, from a software mixing standpoint, was the unique way in which they dealt with mixing and layering sounds. As its author writes:

> When we were working on the IBM [360] mainframe, time was very expensive. It was thus very costly to make mistakes and have to repeatedly redo things. I thus adopted the model of a good rehearsal, where when you make a mistake, you don't go to the beginning and play the whole thing over again, but rather just work on the instruments and the section that are broken. The computer model of this was mixing. I wrote a FORTRAN program called MIX that added its output to the disk rather than simply writing from scratch. Thus, if there were a wrong note in a passage, for example, MIX allowed you to remove it by writing it again with a negative amplitude, thus subtracting it from the mix, and then rewrite it. The mixing approach proved quite useful since it allowed me to save a lot of time and money by just working at it bit by bit. There was no need to redo things that were fine. Thus, the basic difference between MIX and Music-N languages is that MIX had no time sorter, only processed one command at a time, and added its output to the disk. It also allowed one to access data stored on tape at the same time. (Personal communication, 17 July, 2003)

Cmix was more or less a version of Mix written in the C programming language, comprised of a collection of C headers and functions optimized for working with sound. Using these functions, one could write *instruments* and compile them into standalone executable files. A scripting language called MINC, written by student Lars Graf, enabled one to algorithmically invoke instruments in various ways. Cmix has evolved and survives today as a real-time version called RTCmix running on the Linux platform.

Written by Kent Dickey and Paul Lansky, Rt introduced the notion of completely scriptable, automatable software mixing to the DAW. Originally for NeXT computers, it was later ported to Silicon Graphics workstations and allowed users to mix up to 256 tracks of audio. An example mixing script might look something like this:

```
playnote(snd=1,track=1,amp(0,1,4,1,5,0), end=10)
// play sound file 1 for ten seconds with a fade out
playnote(snd=2,track=2,amp(0,0,1,1,2,0), skip=2,end=5,
   transp=-1,at=5)
// play seconds 2-5 of sound file 2, transposed down 1
   semitone starting at time=5 seconds
playnote(snd=3,track=3,at=3,pan=.1)
playnote(snd=3,track=4,at=3.2,pan=.6)
playnote(snd=3,track=3,at=3.4)
```

The notion of the scriptable, text-based mixing of sound files that Rt employed was very powerful. For example, one could write scripts in any programming language or spreadsheet program to batch-process and mix a large number of individual sound files. The tradeoff was that the program had no inherent graphical waveform display of sound files.

Conclusion

Lots of other great open-source and commercial software for mixing and editing sound files has been developed, some of the more recent of which we will discuss in Chapter 4. Once again, owing to its robustness and extensibility, the modern DAW serves as an open canvas for the continual development of new audio software and hardware, making it a powerful and evolving tool ever increasing in its ubiquity and usefulness in the creation and production of music.

For Further Study

"120 Years of Electronic Music." http://www.obsolete.com/120_years/

Andrenacci P, et al. "The New MARS Workstation." *Proceedings of the 1997 International Computer Music Conference*. San Francisco: International Computer Music Association, p. 215–219.

Atkins M, et al. *The Composer's Desktop Project. Proceedings of the 1987 International Computer Music Conference*. San Francisco: International Computer Music Association. 1987; p.146–150.

Borish J. "SoundDroid: A New Approach to Digital Editing and Mixing of Sound." *The BKSTS Journal* 1985;616–621.

Boulanger R. *The Csound Book*. Cambridge, Massachusetts: MIT Press; 2000.

Burns KH. "History of Electronic and Computer Music." http://music.dartmouth.edu/~wowem/electronmedia/music/eamhistory.html.

Buxton W, et al. "An Introduction to the SSSP Digital Synthesizer." *Computer Music Journal* 1978;2(4):28–38.

Buxton W, et al. "The Evolution of the SSSP Score Editing Tools." *Computer Music Journal* 1979;3(4):14–25.

Doornbusch, P. "Computer Sound Synthesis in 1951: The Music of CSIRAC." *Computer Music Journal* 28(1):10–25.

Duesenberry J. "The Yamaha CXSM Music Computer: An Evaluation." *Computer Music Journal* 1985;9(3):39–51.

Freed A. "Recording, Mixing, and Signal Processing on a Personal Computer." *Music and Digital Technology*, 1987;165–172.

Hiller L. "Computer Music." *Scientific American.* 1959;200(6):109–120.

http://www.herbertbrun.org/

http://www.sfu.ca/~truax/pod.html

Jaffe D, Boynton L. "An Overview of the Sound and Music Kits for the NeXT Computer. *Computer Music Journal* reprinted in Pope S.T, ed. The Well-Tempered Object. Cambridge, Massachusetts: MIT Press; 1989;13:(2): 48–55.

Lambert M. "Digital Audio Workstation Milestones," 2001. http://www.prosoundweb.com/recording/articles/mel/daw_milestones1.shtml.

Lansky P. "The Architecture and Musical Logic of Cmix." *Proceedings of the 1990 International Computer Music Conference*. San Francisco: International Computer Music Association, 1990; p. 91–93.

Lindemann E, et al. "The Architecture of the IRCAM Musical Workstation."Computer Music Journal 1991;15(3):41–49.

Lindemann E, Dechelle F, Smith B, Starkier M. "The Architecture of the IRCAM Musical Workstation." *Computer Music Journal* 1991;15(3):41–49.

Lohner H. "The UPIC System: A User's Report." *Computer Music Journal* 1986;10(4):42–49.

Lowe B, Currie R. "Digidesign's Sound Accelerator: Lessons Lived and Learned." *Computer Music Journal* 1989;13(1):36–46.

Loy DG, Abbott C. "Programming Languages for Computer Music Synthesis, Performance, and Composition." *ACM Computing Surveys* 1985;17(2): 235–266.

Manning P. *Electronic and Computer Music*. Oxford: Oxford University Press. 2004.

Mathews MV. "The Digital Computer As a Musical Instrument." *Science* 1963;142(11):553–557.

Moog RA. "The Columbia/Princeton Electronic Music Center: Thirty Years of Exploration in Sound." *Contemporary Keyboard* May 1981.

Moog RA. "The Apple II/Rhodes Chroma Interface." *Keyboard Magazine*, September 1982, 58.

Moore FR. "The CARL Computer Music Workstations: An Overview." *Proceedings of the 1985 International Computer Music Conference*. San Francisco: International Computer Music Association, 1985; p. 5–8.

Pope ST. "The Interim DynaPiano: An Integrated Tool and Instrument for Composers." *Computer Music Journal* 1992;16(3):73–91.

Pope ST. "Computer Music Workstations I Have Known and Loved." *Proceedings of the 1995 International Computer Music Conference*. San Francisco: International Computer Music Association, 1995; p. 127–133.

Scaletti C. "The Kyma/Platypus Computer Music Workstation." *Computer Music Journal* 1989;13(2):23–38.

Schaeffer P. *A la recherché d'une musique concrète*. Paris: Éditions du Seuil; 1952.

Schaeffer P. *Traité des objets musicaux*. Paris: Éditions du Seuil; 1966.

Truax B. "Computer Music Composition: The Polyphonic POD System." *IEEE Computer* 1978;11(8).

Vail M. *Vintage Synthesizers*. Backbeat Books; 2000.

Viara E. "CPOS: A Real-Time Operating System for the IRCAM Musical Workstation." *Computer Music Journal* 1991;15(3):50–57.

Wallraff D. "The DMX-1000 Signal Processing Computer." *Computer Music Journal* 1979;3(4):44–49.

http://www.raymondscott.com/

Exercises and Classroom Discussion

1. Do the tools with which we create music affect the resulting music? Do the tools with which we create music make assumptions about the music we will create with them? Cite specific examples.

2. Find a historically interesting piece of hardware or software that was not discussed in this chapter and present it to the class.

3. Find a picture and description of the *oldest* analog mixing console you can; find a screenshot and description of the *oldest* digital audio workstation software you can. Compare their strengths and weaknesses.

4. Find a picture and description of the most expensive digital mixing console you can; find a screenshot and description of the most expensive digital audio workstation software you can. Compare their strengths and weaknesses.

5. Compare and contrast both historical and current DAW systems based on general-purpose computers alone versus those based on specially constructed hardware in terms of their technical capabilities, robustness, and their lifespan or prospects for long-term survivability.

6. Why is some digital audio workstation software released commercially while other programs are released open-source? What advantages and disadvantages does each method posses?

7. What are the advantages and disadvantages of including dedicated signal-processing hardware (like external processing cards) in a DAW system?

8. In your opinion, what is the most important feature that analog synthesizers have over digital synthesizers? Why?

9. In your opinion, what is the most important feature that the modern DAW offers that the earliest computers used for music-making did not? Why?

Sound Files: Formats, Storage, and Transport

We have spent the first two chapters of this book on a general tour of concepts and an overview of the history of the digital audio workstation (DAW). With a working knowledge of sound, digital audio, and a bit of historical perspective, we can now begin to delve into slightly more technical aspects of the modern DAW. We begin with audio file formats and related terminology, concepts that are essential to working with sound on a modern computer.

Just as graphics and word-processing programs employ either industry-standard or proprietary file formats for storing a user's documents and data, so do DAW programs for editing sound. Whether your DAW is based around Macintosh OS X, Windows, or Linux, it can store audio files in many different file formats. Additionally, the vast majority of modern DAW software is able to write and read many different audio file formats. Among other benefits, this enables users of one platform to easily export their files to another platform. And because portable electronics devices are increasingly able to play sound files of various formats, compatibility with multiple sound file formats can allow such devices to interface with computers and other audio-specific gear.

This chapter begins with an overview of concepts and terminology common to most audio file formats. We then discuss the simplest audio file format, often called "raw" audio, followed by perhaps the two most commonly used uncompressed formats: Microsoft's RIFF WAV format and Apple's AIFF format. We then pay brief homage to less encountered and anachronistic formats for completeness.

In addition to uncompressed audio data, the DAW can of course work with compressed data files as well, such as MP3 and AAC files. We next delve into a brief overview of the technologies involved in these formats, followed by a discussion of the various compressed file formats in common use. And with the proliferation of portable electronics that let us hear audio "on the go," these compressed formats are not going away any time soon!

Related to the specifications of audio file formats is the technology of audio file storage, and so the chapter next addresses the most popular current means of storing digital audio for editing and recall by the DAW: magnetic hard disks, optical and flash storage media, and magnetic tape storage media. We conclude with a listing of ways to cross-integrate multiple DAWs and allow them to communicate with each other, whether they are in the same studio, across the street, or across the world from each other.

You may wonder how much technical knowledge you need regarding audio file formats and storage, and the answer completely depends on your interests. However, those who work at all with audio on computers frequently encounter many of the terms and concepts we will talk about, and the more you know, the better you will be equipped to get the most out of the DAW. There are many file formats from which to choose when working with digital audio, and the specific format chosen can have a great impact on the resulting sound quality and amount of storage space required.

If you are interested in writing your own DAW software, the information presented here is only the tip of the iceberg. Most of the major audio file format specifications are freely available on the Internet, and a lot of free software is available to help the adventurous begin.

Conceptual Overview

Before addressing specific file formats, it is important to understand the relevant basic terminology. A great number of file formats exist, and each one includes its own feature set, strengths, and weaknesses. Because no single format is perfect for all situations, each format attempts to suit a specific target purpose (like small overall file size at the expense of sound quality, or the highest sound quality possible at the expense of larger file size) by trading off some features for others. Regardless of the sound file format, every sound file must tell us something about itself before the DAW can work with it. A conceptual overview of the major components of most sound file formats is shown in Figure 3.1.

Figure 3.1
Conceptual overview of the primary components of most digital audio sound files.

Sound File Headers

How does a piece of audio software know what sound file type it is trying to load? It looks at the *header* of the sound file. The header is a kind of label that indicates the format and specific features of the file. These specific features include the sampling rate of the file, the number of audio channels contained therein, the bit depth (word length) of the samples, and information about the kind of numbers used to encode the audio sample values. (Headerless sound file formats exists, but they are not much used anymore.)

Sound file headers are often enveloped inside a *wrapper* layer, which can include information on whether or not the file is a "streaming" file format (we'll get to that in a bit), licensing and royalty information, copyright notices, and even digital watermarks.

Sampling Rate

We already mentioned the relationship between sampling rate and the highest representable frequency with digital audio. But without the sound file's explicit specification of its encoded sampling rate, DAW software will not know the proper rate at which to play it back, which could result in distortion of its frequency content by shifting all frequencies up or down. (Think of what happens when you spin a vinyl LP faster or slower than its target playback speed. The same thing happens with digital audio.)

The sampling rate is one of the primary determinants of file size. All else being equal, a 192-kHz uncompressed sound file, for example, takes up twice the storage space as a 96-kHz uncompressed file, because it is storing twice as many samples per second (192,000 instead of 96,000).

Number of Channels

Most sound file formats can store mono and stereo audio data, and some can store an unlimited number of channels. When storing more than one channel of data (as with stereo and surround audio files), the channels are generally stored in adjacent *frames*. That is, the sample values for all channels are grouped together before proceeding to the next sample. This processed is called *interleaving*. For example, the frames in a stereo sound file might be organized something like this:

```
<Wrapper/Header>
First frame:  <Channel 1, first sample>    <Channel 2, first sample>
Second frame: <Channel 1, second sample>   <Channel 2, second sample>
 .              .                            .
 .              .                            .
 .              .                            .
Last frame:   <Channel 1, last sample>     <Channel 2, last sample>
<End of File>
```

Occasionally, when working with a DAW, particularly with surround audio files, it may become necessary to split a multichannel file into separate mono files for mixing or processing an individual channel. (However, most recent multitrack mixers support surround audio file formats.) In this case, a sound file editor that supports multichannel audio files can be used to separate the file into separate channels.

Similarly to sampling rate, the number of channels included in a sound file is another determinant of file size. All other things being equal, an 8-channel sound file occupies four times the storage space as a stereo file. Note than when using multichannel sound files, the sampling rate specified in the header corresponds now to the number of *frames* per second, not the number of samples per second. This is done automatically by file-handling systems in the DAW to ensure that each channel of audio plays back at its intended sampling rate.

Bit Depth

Recall that the bit depth (word length) expresses the number of bits used to represent individual audio samples, and therefore determines the number of quantization levels used to encode an analog audio signal. The choice of bit depth used in an audio file, like the choice of sampling rate, can therefore have a tremendous impact on the resulting sound quality.

Support for different bit depths can vary greatly from one format to another. Some file formats, as we will see shortly, support only one bit depth, while others support a wide range of bit depths. In the early days of the DAW, 8-bit and 12-bit bit depths were commonly used due to limited processing power, but modern DAWs most commonly work with 16-bit, 20-bit, or 24-bit samples.

The bit depth is yet a third determinant of file size. All else being equal, a 24-bit sound file occupies 150 percent of the storage space of a 16-bit file (24/16 = 150%). But recall that the 24-bit file exhibits a significantly higher signal-to-error ratio at around 160 dB, versus about 98 dB for a 16-bit file.

Fixed Point versus Floating Point

Audio samples can be stored on a computer using either fixed-point or floating-point representations. A fixed-point number assumes that the location of the decimal point stays constant. With audio samples expressed as fixed-point numbers, the decimal point is assumed to be located directly after the first (leftmost) bit position. (We will say more about this in a moment.)

However, the floating-point representation allows the radix to freely move, provided we (or our software) keep track of it. (Remember scientific notation?) This is particularly useful for audio, because it means that each sample can always take advantage of the full dynamic range of the number of bits available. That is, floating-point numbers can express smaller numbers with the same numerical precision that they can larger numbers. This is not true of fixed-point numbers: As the magnitude of the number they are expressing decreases, so does the number of bits they use to express the number.

Whether a sound file's samples are stored as fixed-point or floating-point numbers can significantly affect the sound quality, particularly after a sound file is processed. Consider a simple example: Two sound files, each recordings of the same source, are identical except that one uses fixed-point samples and the other uses floating-point samples. Examine the difference after significantly attenuating each file's amplitude in a sound file editor and then amplifying them back to their original levels, shown in Figure 3.2.

Multitrack editors generally mix and process sound files internally using high-precision 32-bit (or higher) floating-point arithmetic. This is done to minimize the overflow and round-off errors that are likely to occur when mixing and processing multiple sound files together. (Note that to mix and process fixed-point sound files together, the software must internally convert the files' individual samples to floating point values.)

Figure 3.2
Attenuating and
subsequently
amplifying sound
files with fixed-
point samples and
floating-point
samples.
(a) Time-domain
representation of
original fixed-point
sample sound file;
(b) original
floating-point
sample sound file;

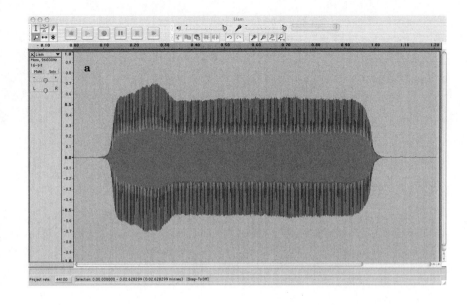

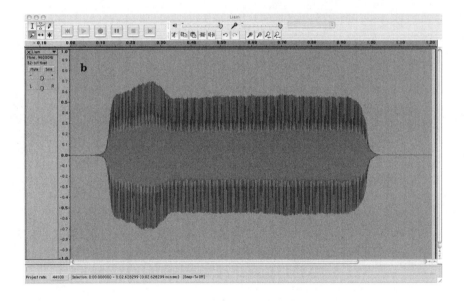

Figure 3.2
Attenuating and subsequently amplifying sound files with fixed-point samples and floating-point samples. (c) attenuated fixed-point sound file; (d) attenuated floating-point sound file;

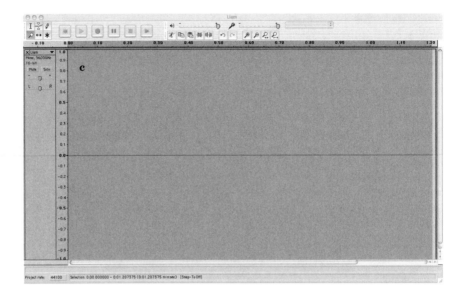

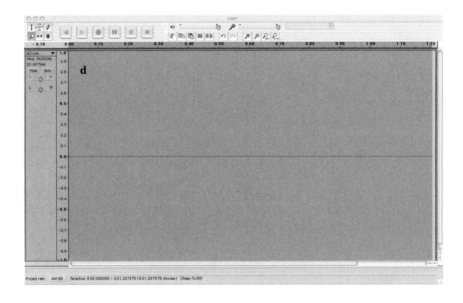

Figure 3.2
Attenuating and subsequently amplifying sound files with fixed-point samples and floating-point samples. (e) after amplifying the attenuated fixed-point file (note the increased noise floor); (f) after amplifying the attenuated floating-point file (note this is the same as (b)).

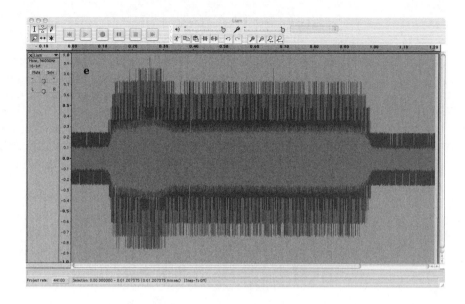

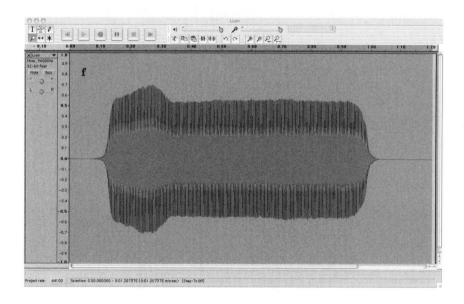

Signed, Unsigned

Likewise, samples can be encoded as *signed* or *unsigned*. This applies to both fixed-point and floating-point numbers. Unsigned numbers range from a value of 0 up to the highest number expressible with the given number of bits; signed numbers range from the negative of half the maximum possible value up to the positive of half the maximum value.[1] Our DAW software needs to know whether the sample values it is dealing with are encoded as signed numbers or as unsigned numbers so that it interprets them correctly.

Different methods can be used to denote positive and negative sample values. The simplest is by using a sign bit, usually the leftmost bit, where a 0 denotes a positive number and a 1 denotes a negative number. For example, $0001_2 = +1$, taking the leftmost zero to be a positive sign bit, whereas $1001_2 = -1$ since the leftmost 1 indicates this is a negative number. (Note that we lose one bit of resolution when using a sign bit.) As an unsigned number, 0001_2 would still represent a value of $+1$, but 1001_2 would represent $1 \times 2^3 + 0 \times 2^2 + 0 \times 2^1 + 1 \times 2^0 = 8 + 2 = 10$.

Another way that sound files can denote signed integers is a notation known as *two's complement*. Thought of as a mathematical operator, taking the two's complement of a number yields the negative of its value. It is simply found by inverting each bit of a binary number and adding one to the result. For example, let's take the 4-bit number $0011_2 = 3$. To represent the inverse of this number, that is a decimal value of -3, we invert the bits, yielding 1100_2, and add 1, for a result of $1101_2 = -3$ in two's complement notation. Note that the same process can be used to convert backward and forward between positive and negative numbers: We still just invert the bits and add one.

Bit Order

The terms *big-endian* and *little-endian* are used to describe the ordering of bits used to represent sample values. The terms are taken from Jonathan Swift's 1726 novel *Gulliver's Travels*, in which they denoted which end of a hard-boiled egg should be cracked first. The endianess of a number or word determines the order in which we should read it;

---

1 Actually, because the number zero is neither positive or negative, we typically restrict signed representations to be in the range from the negative of half the maximum possible to the positive of half the maximum, minus one.

applied to addresses on the World Wide Web, this would mean the difference between www.mcgraw-hill.com and com.mcgraw-hill.www.

Big-endian representation means that the most-significant bit—the "big end"—of the number is stored first, in the lowest computer memory address. This is the way we typically read numbers, from left to right, with the most-significant bit at the leftmost part of the number. Little-endian representation means that the least-significant bit—the "little end"—of the number is stored in the first, leftmost bit position. Computer processors vary in their use of big-endian or little-endian representations: PCs typically use little-endian, while Macs can use either. Endianness does not affect sound quality, and it is usually not anything worth worrying about, unless you are writing software for the DAW.

Companding

Some sound files allow analog signals to be quantized and stored along a logarithmic, rather than a linear, basis. By spacing lower-amplitude quantization levels closer together and higher-amplitude levels farther apart (see Figure 3.3), the apparent dynamic range can be extended. This process is called *companding*, and two kinds of quantization-level spacing techniques are commonly used: μ-Law ("myoo-law") and A-Law. The result is that the lowest possible quantization level of an 16-bit μ-Law signal is roughly the same as a 20-bit linear (noncompanded) signal. The idea behind companding is that the majority of samples in most audio signals stay relatively low in value, only occasionally approaching the maximum possible value, so a larger signal-to-error ratio can be achieved most of the time. Note that companding does not affect file size.

Data Compression

To reduce the amount of space needed to store sound files, various data compression techniques can be used. Some sound files support compressed data, while others do not. Data compression can be either *lossy* or *nonlossy* in nature. A lossy compression scheme, like MP3, can achieve an enormous reduction in file size (like 10:1), but at the expense of not being able to perfectly reconstruct the original, uncompressed signal. Nonlossy compression schemes trade off file-size reduction rations for the ability to perfectly recreate the original sample values.

Figure 3.3
Companding
algorithms
nonlinearly
distribute the
available
quantization
levels. (a) One
period of a sine
wave; (b) sine
wave after 4-bit
linear quantization
(no companding);

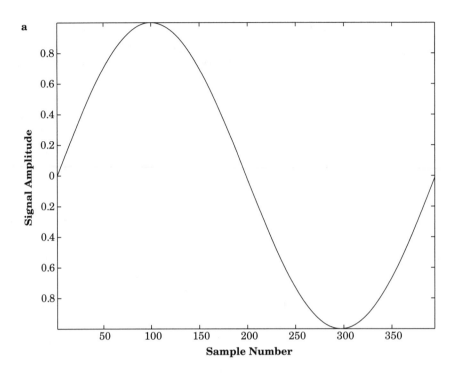

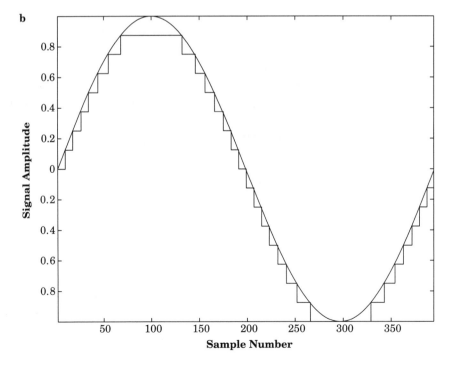

Figure 3.3
Companding
algorithms
nonlinearly
distribute the
available
quantization
levels. (c) sine
wave after 4-bit
logarithmic
quantization
(μ-Law
compression); (d)
sine wave after
μ-Law expanding.
The original sine
wave is overlaid in
each figure for
comparison.

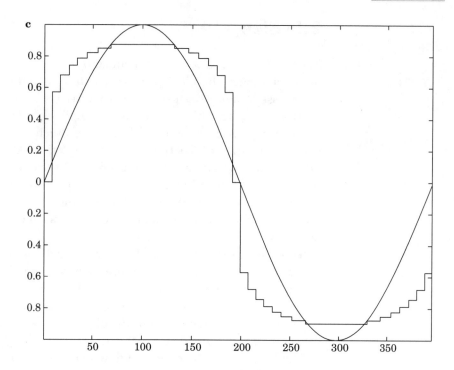

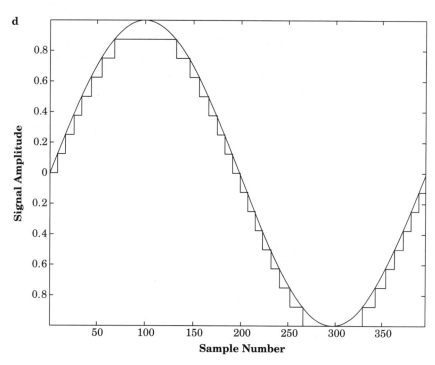

File Size

The size of a sound file is determined by the sampling rate, the number of channels, the bit depth of the samples, and whether the file is compressed (and if so, how it was compressed). (The header takes up a little space, but this is generally negligible.) File sizes can be computed easily using dimensional analysis. Each minute of an uncompressed 16-bit, 44.1-kHz stereo file ("CD-quality") occupies 16 bits/sample x 44,100 samples/second $\times$ 60 seconds/minute $\times$ 2 channels = 84.672 megabytes (Mb). By comparison, each minute of an uncompressed 24-bit, 192-kHz stereo sound file occupies 24 bits/sample $\times$ 192,000 samples/second $\times$ 60 seconds/sample $\times$ 2 channels = 552.96 Mb.

To compute the file size of a compressed sound file, we need to know something about the compression scheme that was used. For the purposes of file size, all we need to know is the *compression ratio* or the *bit rate* of the compression algorithm used. For example, a 10:1 compression ratio means that 10 Mb of data was compressed into 1 Mb, or 1/10 the original size.

The bit rate, on the other hand, is more often encountered. It quite simply tells the number of kilobits per second of audio (thousands of bits per second) the file occupies. For example, a 128-kbit/sec MP3 file occupies 128,000 bits for each second of the file. To convert this to kilo*bytes* (kB) per second, just recall that there are 8 bits in a byte, so simply divide the bit rate by 8. Thus, our 128-kbit/sec MP3 file occupies 16 kB of storage space per second. The term *low-bit-rate coding*, for which the primary goal is the development of high-quality audio compression representations that consume the minimum number of data bits possible, is often encountered, and we will talk more about it later.

File Conversion

A sound file converter is a program used to convert sound files from one format into another. This functionality is incorporated into most DAW software for commonly encountered formats, but a standalone converter program might be needed when converting to or from a more obscure format. It is important to be aware of the characteristics of the source and target file formats to prevent unintentional loss of data, for example, by converting from a 192-kHz, 24-bit file to a 44.1-kHz, 16-bit file. This may be what you want to do in some cases, but knowing the basic features of various file formats can help prevent mistakes.

Uncompressed (Usually) Formats

As mentioned, sound files can be either compressed or uncompressed in nature. We first examine each of the uncompressed sound file formats in common use, followed by the popular compressed formats. For completeness, we then consider two other kinds of audio formats frequently encountered on the modern DAW: multitrack audio file formats, which allow multitrack mixes to be saved in one mixer and opened in another, and looping formats, which are specifically designed to be used with loop-editing software.

Headerless Audio Files

The most basic uncompressed format is the headerless audio file, often called "raw" audio. Raw audio files are simply text or binary files that contain a stream of sample values. Because they are headerless, it's an easy hack for any software you write to generate them: Just write sample values to a plain text file. But this simplicity comes at a price: What's the sampling rate, the number of channels, the bit depth? The file extension can be interpreted to answer these questions, but it gets confusing. For this reason, headerless audio files are not often used, but they are occasionally encountered.

Audio Interchange File Format

The Audio Interchange File Format (.aiff or .aif) is the format most often encountered on the Macintosh. Developed by Apple Computer under the general umbrella of the Electronic Arts Interchange File Format (called the EA IFF 85 Standard), it is used to store high-quality uncompressed samples. (An extension to the AIFF format that provides for compression, called AIFF-C [.aifc] also exists, although it is not as commonly used.)

AIFF files are very flexible in that they support arbitrary sampling rates and bit depths as well as any number of audio channels. The specification also supports companding as well as both fixed- and floating-point samples.

The Interchange File Format calls for data to be stored in chunks of varying types, which can allow a wide range of information to be stored along with audio data. Notably, AIFF files can include *markers*, which are like "bookmarks" to specific places within the file. This can be especially useful for marking key locations in a long file, for quickly jumping back and forth between them while editing, or for defining looping points. In fact, markers can be used for any purpose an audio application desires; they are just placeholders, and it is up to the application to determine how to interpret them. The AIFF specification also allows other data, like playlists, text comments, and MIDI messages, to be stored within the file as well.

Wave

The Wave (.wav) format, jointly developed by Microsoft and IBM, is similar to the AIFF format in that it is based on the general Interchange File Format specification and supports a wide variety of sampling rates, bit depths, companding, and an arbitrary number of channels. Eight-bit sample values are stored as unsigned integers, whereas 16-bit sample values are stored as two's-complement signed numbers. The specification is not quite as elegant as AIFF from a programming perspective, but this does not affect the sound quality. Wave files are the uncompressed format in common use on Windows operating systems. In general, Wave and AIFF files can be freely converted between each other without loss of data.

The Wave format does allow compressed sample data to be stored, but Wave files are generally used in professional DAW applications only for storing uncompressed data. A problem with the compression of Wave files is that so many different compression schemes can be used that it can make writing a program to deal with all of them a bit tedious. Because the Wave format is based on the Interchange File Format, it too allows chunks to store different kinds of data within the audio file.

Miscellaneous Uncompressed Formats

AIFF and Wave are by no means the only sound file formats primarily used for storing uncompressed audio; a great many others exist. The two that are probably most frequently still encountered are formats developed back in the 1980s: Sound Designer II and NeXT/Sun.

Sound Designer II format. Digidesign developed the Macintosh-specific Sound Designer II (.sd2) format to accompany its popular Sound Designer II two-track editing software in 1985. Although this program is no longer available, the format was picked up by other software, and SD2 files are still occasionally encountered. The specification supports a wide range of sampling rates and bit depths, as well as an arbitrary number of channels. Data is stored as signed integers, and companding is not allowed.

The NeXT/Sun format. The same file format structure was used by both NeXT and Sun computers of yesteryear. Using the suffix ".snd" on the NeXT and ".au" on Suns, the two file types are more or less interchangeable. The NeXT/Sun Format supports a large number of sampling rates and bit depths as well as companding, signed integer sample values, and floating-point sample values.

...And the rest. The remaining uncompressed audio file formats can in general be broken down according to the community in which they were developed. The speech and telephony research community developed its own, the computer music research community developed its own, and the consumer/prosumer audio industry developed yet others. Many of these formats are still in use by their respective communities, but thankfully AIFF and Wave have emerged as the de facto standard for storing uncompressed, professional-quality audio data on the DAW.

If you do ever encounter an unrecognized audio file format (or a file extension of any kind, for that matter), the Internet contains many file format repository sites for further information. And several stand-alone file format converter programs are available (many of them free) that can translate even many of the more obscure formats into AIFF or Wave files.

Compressed Formats

As mentioned, compressed audio formats can be lumped into two groups: lossy and nonlossy. Before we discuss specific compressed file formats, it is important to mention the separate but related concepts of *compression scheme* and *codec*. A compression scheme is a general algorithm, or series of steps, used to reduce the size of computer file.

A codec, or encoder/decoder, is a specific implementation of a compression scheme. Many codecs are available for different compression schemes; some of them are free and open-source, while others are commercial products.

Codecs can further be classified according to whether or not they support streaming. When a sound file is streamed over its distribution channel (whether the Internet, satellite television, etc.), the receiving device stores incoming data into a memory buffer. Once the buffer begins to fill up, the audio is decoded and played back. Streaming is particularly useful if the distribution channel cannot maintain a constant bandwidth, as is the case with the Internet and satellite channels.

How should one choose which codec to use? The choice between a lossy and nonlossy codec is often made on the simple trade-off consideration of file size versus sound quality. This trade-off has been blurred considerably in recent years, however, with the advent of high-ratio compression codecs that can fool listeners in blind tests. With the influx of new technologies like satellite radio, Internet radio, and portable audio devices that can store hundreds of hours of music, the choice is often made on the basis of bandwidth, and most consumers of music seem quite content with the lossy codecs currently available.

Choices among codecs are often made based on the intended distribution channel. Compact discs, DVD-Video, DVD-Audio, and Super Audio CD (SACD) each support different codecs. Still other choices among codecs are made on the basis of cost.

Compressed file formats are generally not used until a mix is complete. Once the uncompressed files have been processed, mixed together, and mastered, the choice of a codec can then be made.

Lossy Compressed Formats

Perhaps the most commonly encountered lossy compressed formats are MP3 and AAC. Many others also exist—more than we can possibly mention here—and many of them are proprietary formats developed by one company for use only by that company. Thankfully, many great DAW programs can automatically encode sound files into the more popular lossy compressed formats. However, care should be taken not to convert one lossy compressed format into another if possible; it is far better to convert the original, uncompressed file into a lossy compressed format rather than impart multigenerational degradation to a sound file.

If we are going to accept losing data to gain a greater data reduction, how do we choose what parts of the audio signal to discard? Most lossy compression schemes are based on a mathematical model of human psychoacoustics. One of the primary ideas behind these models is that we are not able to accurately discern frequencies that have been *masked* behind other neighboring frequencies or by noise. Think of a somewhat analogous scenario in three-dimensional computer graphics: If an object is in front of (masking) another object behind it, there is no need for the computer to render the masked object; it would be a waste of processor resources, because we could not see it anyway. The computer just needs to keep track of where the masked object is in case it becomes no longer masked, and hence visible. Other psychacoustic effects are also taken into account, and as we continue to learn more about human psychoacoustics, surely even better compression schemes can be forged. The MPEG 1 specification even provides a couple of different psychacoustic models, and some encoders allow the user to specify which model to use.

Lossy compression codecs generally support multiple bit rates in which a file can be encoded. A file encoded at a low bit-rate, for example 28 kbps, exhibits lower file size (or required streaming bandwidth) and lower overall sound quality than the same file encoded at a high bit-rate, like 160 kbps. To complicate matters, some encoders allow the user to specify constant bit-rate (CBR) or variable bit-rate (VBR) encoding.

Note that the bit rate is distinct from the sampling rate. Uncompressed audio in any sampling rate can be converted into any bit rate. The bit rate, in a practical sense, just specifies how "severe" the compression will be by limiting the number of bits per second that can be used to represent and later reconstruct the audio file.

MPEG audio. The Motion Picture Experts Group (MPEG) is an international steering committee that was formed to develop and recommend compression standards for digital audio and video. MPEG represents a collection of specifications (some of them free and open, others not) that attempt to standardize and codify compression schemes with the International Standards Organization (ISO); the most well-known is perhaps the MPEG 1-Layer 3 scheme, commonly called MP3 (.mp3). (A "layer" is a kind of subcategory in the specification.) Other MPEG compression schemes are available and continually being developed by the international research community; collectively, these algorithms strive for the low-bit-rate coding of digital multimedia content.

MP3. Although the MP3 compression scheme was codified by the MPEG Working Group, the actual technology on which the standard is based was jointly developed by two companies: the Fraunhofer-Gessellschaft (http://www.fraunhofer.de) and Thomson (http://www.thomson.net), who hold several patents in this area. To write an MP3-compliant codec, a fairly hefty licensing fee must be paid. And any time MP3-encoded content is sold, you must technically pay a royalty.

One of the nice features about MP3, aside from its ubiquity, is its support for meta-data. As online music stores and distribution channels have proliferated, so too proportionately has grown the ability to "tag" multimedia files with information about their contents.[2] This can allow search engines to quickly locate files that match specific search criteria.

Incidentally, MPEG-1 Layer 2 (.mp2) files are still occasionally encountered. They sound great, and files can typically be encoded in MP2 format faster than in MP3 format, although the resulting MP2 files tend to be larger.

Many MP3 codecs are now available. One particular item worth mentioning is LAME (a recursive abbreviation for "LAME Ain't an MP3 Encoder"), available online at http://lame.sourceforge.net. LAME began as a hack around the original ISO MP3 specification to produce royalty-free MP3 files, but it is now a complete, standalone encoder that does not use the ISO source but produces MP3-compliant bit streams. LAME functionality has been incorporated into many DAW programs for Linux, Windows, and Macintosh operating systems.

AAC. The MPEG Working Group did not stop with the MPEG 1 specification; the MPEG-2 specification also includes a Layer 3 specifically intended to work with lower sampling-rate audio files. MPEG-3 was rolled back into the MPEG-2 specification, and so we jump over next to MPEG-4.

MPEG-4 represents another broad umbrella of next-generation multimedia compression schemes. The audio portion (MPEG-4, Subpart 4) encompasses two subsections: AAC, developed by a group of

---

[2] In fact, the concept of *music information retrieval,* which can allow audio databases to be queried by humming a portion of tune, or by requesting a song that sounds like another, even without tags, has fostered an international research community (see http://www.ismir.net).

companies including Fraunhofer, Dolby, Lucent, and Sony; and VQF, developed by Yamaha and Sony. Both VQF and AAC files can be found on the Internet, although AAC is much more common. In fact, AAC seems to have won the battle as the successor to MP3 technology. It is said that both AAC and VQF files sound much better than MP3 files encoded at the same bit rate; in fact, comparable audio quality can be achieved with AAC and VQF at almost half the bit rate of an MP3 file.

MPEG-4 treats audio and video in a completely object-oriented manner; that is, it defines everything it encodes as an object that is potentially malleable by the decoder. Consider its support for *structured audio*, for example. Structured audio describes the kind of audio to be produced (synthesized) by the decoder. For example, rather than encoding a recording of a violin, with structured audio we could just write instructions for how and when to synthesize it, which can save an enormous amount of storage space and bandwidth. The structured-audio specification is based on the Csound sound synthesis environment, a direct descendant of the original Music-N environments developed at Bell Labs beginning in the 1950s. MPEG-4 also supports multichannel audio, so surround mixes can be encoded and decoded.

Other MPEG formats are just over the horizon. MPEG-7, still under consideration, calls for objects to be represented in eXtensible Markup Language (XML) for even greater interoperability and indexing capabilities. And the planned MPEG-21 envisions a robust multimedia content distribution mechanism that works on computers, cell phones, televisions, and PDAs and incorporates additional intellectual property management. Indeed, as digital information becomes ubiquitous, the ongoing quest for approaches to digital rights management and protection of intellectual property will only continue.

Ogg Vorbis.   Ogg Vorbis (http://www.vorbis.com) is a relative newcomer. It is unique among non–MPEG-based codecs in that it is free, it is open-source, and is not patented. It is often cited as sounding as good or even better than many commercial codecs. And because no license is needed to use the encoder, no royalties must be paid. The Ogg Vorbis (.ogg) format is used for Internet audio streaming and files, as well as for game audio. The specification has fostered quite a large user base and support community.

ATRAC.   The Sony MiniDisc uses its own kind of low-bit-rate encoding called Adaptive Transform Acoustic Coding (ATRAC). Like current MPEG audio technology, known psychoacoustic principles are used to eliminate perceptually redundant or imperceptible audio information, and the ATRAC algorithm can achieve compression rates of about 5:1. Next-generation algorithms based on this technology include Sony's ATRAC3 and ATRAC3plus, which can yield 10:1 and 20:1 data compression rates, respectively, and are intended for use primarily with portable audio electronics devices and over the Internet.

WMA. Microsoft's Windows Media Player format, dubbed Windows Media Audio (.wma, .asf), is often encountered. It is mainly used for very low bit-rate encoding to minimize downloading times over the Internet and is a subset of Microsoft's Advanced Streaming Format (ASF) standard.

RealAudio. The RealAudio (.ra, .ram) format is a proprietary technology owned by RealNetworks and used mostly in their suite of streaming tools for Internet audio broadcasting. It is perhaps the main corporate competitor of the WMA format, and it seems to have found a particular niche among Internet radio (and now video) broadcasters.

...And the rest. Other lossy compressed audio formats are often encountered, including Quicktime, Liquid Audio, and others.

Quicktime is a collection of codec technologies from Apple Computer. Used primarily in the simultaneous encoding of audio-visual material, it can be used solely for audio files and supports both downloading and streaming. The technology is available for both Macintosh computers and PCs.

LiquidAudio encompasses a suite of technologies built around an electronic-commerce music distribution model. The format allows digital audio files to be bought and sold over a secure server, with payments automatically distributed to the appropriate parties.

A large number of other lossy compression formats exist, particularly in low–bit-rate applications like speech and telephony. More information about these is available on the Internet.

Lossless Compressed Formats

Lossless compressed formats are frequently used on all kinds of computer files. Popular lossless compression schemes like ZIP and SIT can be used on audio files, but, with the exception of the newer RAR lossless scheme, much better compression ratios can be achieved when using an algorithm specifically written to work with audio.

Lossless encoding is used to compress digital audio in such a way that the original, uncompressed file can be perfectly reproduced. These compression schemes have advanced a great deal in recent years, and file sizes can now easily be cut in half or even a third or more without loss of any data. While not currently as popular from a music distribution standpoint as lossy formats, lossless compression formats are particularly appealing for users wishing to compress large libraries of audio. This can be especially useful for archiving gigabytes of sound samples and sound effects for later recall in a mixing environment.

Like their lossy counterparts, many lossless compression schemes and encoders are available, and new ones are continually being developed. Most of them work by searching for data redundancies, replacing them with shorthand notation, or codes, that can later be expanded. Data can also be compressed by running a number-crunching algorithm to attempt to predict each subsequent sample value and storing only the error between the prediction and the actual value. Other cool tricks can cut down file sizes even further.

FLAC.   FLAC (http://flac.sourceforge.net), the Free Lossless Audio Coder, is a very expensive encoder. (Okay, I just wanted to make sure you were still awake!) Its cross-platform support, fast decoding speed, and open source make it particularly appealing.

WMA.   Windows Media Audio (.wma) files can also be encoded losslessly. A simple setting on the WMA encoder lets the user select lossy or lossless encoding.

MLP and others.   Many other lossless compression formats exist. Perhaps the most commonly encountered currently is Meridian Lossless Packing (MLP), which is supported by the DVD-Audio standard and can achieve compression ratios of around 2:1. Other formats include Monkey's Audio (APE), Lossless Audio (LA), Linear Prediction Audio Codec (LPAC), OptimFROG, RK Audio (RKAU), Shorten (SHN), String Sort ZIP (SZIP), and WavPack.

Miscellaneous Audio File Formats

In addition to the compressed and uncompressed file formats we have talked about so far, other audio file formats have recently been developed that are tailored for working with a specific application. These include looping formats for use with loop-editing DAW software and multitrack formats that can allow the transfer of entire mixes among computers and different software.

Before we consider looping and multitrack file formats, however, we should mention the idea of representational file formats, like MIDI and MOD. The Musical Instrument Digital Interface (MIDI) specifies a file format (.mid) for storing representations of musical events. It is not an audio file format, however, but it is still much in use. In fact, we will talk much more about it in a future chapter. MOD (.mod) files, although they store audio samples, are not a professional audio file format. They are a kind of hybrid representational/audio format that stores recordings of instruments along with instructions about when they should be triggered.

Sampler File Formats

Hardware-based samplers, which can load, edit, store, and play back digital audio, have largely been replaced by software running on the DAW. Before clear audio file formats emerged, the manufacturers of these hardware samplers developed their own file formats for storing recordings to be loaded and stored in a sampler. These formats, like AKAI S6000, ASR10, SampleTank, SampleCell, EXS24, and LM4 have been more or less superseded by standard AIFF and WAV formats, which most modern samplers (both hardware and software) recognize. However, some DAW-based samplers do still use proprietary sample formats.

Looping Formats

The many loop-based mixing programs currently available include their own specifications for storing audio loops in a way that they can easily be used by the host. Apple Loops is one such format to which AIFF files can easily be converted. The Apple Loops format simply embeds meta-data regarding information like loop points, tempo, and key signature into the AIFF data tags. Other looping file formats exist, like Propellerhead Software's format for its ReCycle software (.rex2) and Sony Multimedia's format for its ACID program. Just as Apple Loops embeds meta-data into an AIFF file, the ACID format typically embeds meta-data into a Wave file. Such an embedded Wave file is said to be "Acidized."

Multitrack Formats

Most multitrack editor/sequencers (like Pro Tools, Sonar, Cubase, and Digital Performer) use their own proprietary file formats for storing all the mix parameters, track settings, plug-in preferences, and so on for a particular mix session. Using these files, users can save their work and return to continue the mix at a later time. However, because these formats are proprietary, most of them are not interchangeable.

The idea of a universal multitrack file format is a noble one. One could create a mix using one piece of software on a Macintosh, save it on a DVD-R, and then mail it to a mastering engineer who would then master the mix on a different piece of software on a Windows computer, for example. All of the individual tracks, plug-ins, and settings would be preserved. Because there is not a de facto standard multitrack mixer for all situations and budgets, many DAW users could benefit from the ability to freely interchange mixes among computers and operating systems.

In 1992, such an attempt was made. Avid Technology (which now owns Digidesign) offered the Open-Media Framework Interchange (OMFI). Several multitrack mixers supported (and still do to some extent) this format, which could be used to transport all kinds of media (not just audio) files among software programs, but OMFI is no longer developed. For various reasons, it became the source of much frustration among DAW users, and it never fully caught on as a universal standard.

Other technologies may emerge to fill the void. One of these is the Advanced Authoring Format (AAF; http://aafassociation.org and http://aaf.sourceforge.net), an XML-based media format developed by a broad consortium of companies. Another viable option may lay in future MPEG specifications.

Summary of Major Digital Audio File Formats

We've examined a wide array of digital audio file formats. However, most DAW users typically only encounter a small handful of formats. The features of these commonly used formats are summarized in Table 3.1.

TABLE 3.1

Summary of Major Digital Audio File Formats

Format	File Extension	Compressed?	Multichannel (>2) Support?	Multitrack Support
Raw	Varies	No	Yes	No
AIFF	.aif, .aiff	No	Yes	No
AIFF-C	.aifc	Yes	Yes	No
WAV	.wav	No, although it can be	Yes	No
SD2	.sd2	No	No	No
OMFI	.omf	No	Yes	Yes
MPEG-2	.mp2	Yes	No	No
MP3	.mp3	Yes	Sort of[1]	No
MPEG-4	N/A[2]	Yes	Yes	Yes
MPEG-7	N/A[3]	Yes	Yes	Yes
Ogg Vorbis	.ogg	Yes	Yes	No
AAC	.aac	Yes	Yes	No
ATRAC	.atrac	Yes	Yes	No
WMA	.wma	Yes	Not currently	No
RA	.ra, .ram	Yes	Not currently	No
MLP	.mlp?	Yes, but non-lossy	Yes	No

[1] At the time of this writing, a multichannel MP3 encoding is not yet available, but it has been proposed.
[2,3] File extensions for MPEG-4 and MPEG-7 files have not been codified at the time of this writing.

Overview of File Storage

When working on the DAW, you are likely to encounter many different audio file formats. Likewise, many choices are available for storing these files, either for archiving or distribution, and each has its own strengths and benefits.

Hard Disk Storage

Hard disks are of course standard on any modern computer, and they are the primary storage medium for audio files. Hard disks are fast enough, and data access times are low enough, that many tracks of audio can now be processed and mixed without ever leaving your computer's internal hard disk. This wasn't always the case with the DAW; in the early stages of its development, specialized disk interfaces and expensive hard disks were required. Thankfully, hard disks are now a dime a dozen, and most DAWs benefit from the largest, fastest drive possible. It is a good idea, though, to regularly defragment your DAW's hard disk, which can make it run much more efficiently.

Hard disks primarily are used for storing, archiving, and working with sound files. However, the many tiny, portable drives that are now available have made them quite popular for transporting data from one machine to another as well.

Optical Storage

Optical storage media, such as compact discs and DVDs, are used for archiving, transport, and distribution. It offers one of the least expensive storage options. In fact, blank compact discs can often be had for free or next to free (thanks to rebates), and blank DVDs seem to be headed that way.

CD Compact-disc recordable (CD-R) drives are increasingly ubiquitous and are in fact built into most recent computers. Coupled with the negligible price of blank CDs, the compact disc has become a more-or-less give-away product: One can produce and archive or give away compact discs gratis.

Compact discs can be formatted to store data (data CD-Rs) in any format, or they can store audio in a format that can be recognized by commercial CD players (in either uncompressed stereo or multichannel DTS format). The compact disc can store approximately 640 Mb of data, or about 74 minutes of uncompressed 16-bit, 44.1-kHz audio.

Super Audio Compact Disc (SACD) is a relatively new distribution format from Sony that supports high-definition audio (with a frequency response of over 100 kHz and a dynamic range over 120 dB). Although not used for file storage or archiving, SACDs can hold up to six discrete channels of uncompressed audio for distributing stereo and surround mixes.

DVD.   DVD is officially not an acronym; it is simply three letters. Some of the people behind the standard originally proposed "Digital Video Disc," but the medium is not restricted to storing video; the phrase "Digital Versatile Disc" was later proposed but never officially adopted. DVDs can store much more data than compact discs: about 4.7 gigabytes using only one side and a single layer. (Up to two physical layers can be used on one or both sides.) Two standards are available for producing, storing, and distributing audio on a DVD: DVD-Video and DVD-Audio.

As far as the DAW is concerned, DVD serves four main purposes: storage, archiving, transport, and distribution. The DVD-Recordable (DVD-R) format can hold up to 4.7 Gb of data, room for plenty of audio (and video) files. Audio produced on the DAW can be distributed on DVD-Video format or DVD-Audio format. We talk more about these formats, as well as SACD, in the final chapter of this book on mastering and distribution.

Flash Storage

Flash storage devices have become increasingly popular. Both inexpensive and portable, these units can be as small as a key ring and

Figure 3.4
High-capacity compact flash cards (shown) and flash pen drives can hold hours of high-quality audio data.

yet hold hundreds of megabytes of data. (See Figure 3.4.) Inside, data is stored in a special electrically erasable programmable read-only memory (EEPROM) that can quickly erase and store data...in a flash!

Digital Magnetic Tape Storage

Although still used in some studios, magnetic tape storage is falling out of favor. Prized for their ability to store large amounts of audio data on relatively inexpensive tapes, digital tape recorders (DTRs) have been the mainstay of multitrack recording facilities for many years; some also use them to back up audio from their DAW. They have been most used, however, for multitrack recording in a studio for later editing on a DAW.

However, perhaps owing to lower cost, portability, and widespread acceptance of the next generation of storage media, coupled with the DAW's direct-to-disk recording capabilities, many new studios based on the DAW are finding less need to purchase a separate DTR, but we mention them here for completeness.

DAT. The popularity of Digital Audio Tape (DAT) has been largely due to the role it has served in portable, in-the-field stereo recording devices. With a battery-powered DAT recorder, headphones, and a microphone, any sound could be recorded at 44.1kHz or 48 kHz. Some people even rigged their DAT recorders with 12-volt car batteries so that they could record out in the field for days at a time! As far as on-the-spot recording, these devices have, however, largely been replaced recently by portable flash recorders, CD- and DVD-R recorders, and direct-to-disk laptop recording.

Data files, rather than audio files, can be stored to Data DAT recorders, and for many years, Data DAT devices have been used as a daily computer backup solution. Again, though, they have been largely replaced in this capacity as well.

ADAT. The widespread acceptance of the format was caused by two of its key features: low cost, and ability to record up to eight channels of audio at a time. Plus, ADAT recorders use readily available Super-VHS (SVHS) tapes as their recording medium, and multiple units can be chained together to record more channels. ADAT uses an optical cable (called a "Lightpipe") that can transmit and receive eight channels of audio at a time, including word clock synchronization information (which we will discuss in Chapter 9). As such, many audio interfaces for DAWs feature ADAT Lightpipe connections. Although not able to store data audio files per se, Alesis Digital Audio Tape (ADAT) has been used to back up and store digital audio data, but this is not an ideal solution.

DA-38/DA-88/PCM-800.   The Tascam DA-88 and Sony PCM-800 DTRs are similar units that record digital audio to Hi-8 Videocassettes, but again not as data files. Both these DTRs use the Tascam Digital Interchange Format (TDIF) to send and receive up to eight channels simultaneously over a 25-pin cable, but word clock must be sent separately.

Cross-Studio Transport and Integration

At some point in the life of the DAW, it becomes necessary to transfer audio from one computer to another, either for disc replication, mastering, using software that is only available for another operating system, or simply sharing music with friends. If frequent file transfer is desired, for example if your studio contains two or more computers, a more permanent solution may be needed.

CD/DVD

One of the easiest ways to quickly transfer files from one computer to another is by simply burning a CD or DVD. It seems that these media have already replaced the floppy disc of yesteryear, as they have become more or less disposable.

Internet

For simply transferring individual files that are not too large, electronic mail may be a good choice. (Some of us even mail files to ourselves quite often for transfer or backup.) For larger files, it may better to place them on an FTP or Web server.

Portable Drives

Portable hard drives with Universal Serial Bus (USB) or the faster IEEE 1394 ("FireWire") interfaces represent yet another popular choice for transferring files among computers. For that matter, if your studio is based on a laptop, that too can serve as the perfect file transfer device. Most operating systems allow laptops to be rebooted in "hard drive" mode so that they appear as a portable hard disk when plugged into another computer.

Removable Media

Many larger and multiuser DAW-based studios employ a more semipermanent solution of removable media and swappable disk drives. These hardware devices remain plugged into the DAW, and yet individual users can purchase their own high-capacity removable media and freely swap them to and from the drive bay. Smaller amounts of data can also be stored and transferred using flash memory cards and card readers.

More Permanent Solutions

For multicomputer DAW studios in which files are frequently transferred among computers, a more permanent, "hardwired" solution is desirable. Their infrastructure can cost a bit more to implement, but they save tremendous amounts of time and are more robust.

Networking.   The least expensive option is to simply network multiple computers with a Local Area Network (LAN). With an inexpensive router, computers can quickly be connected together, optionally sharing a single Internet connection. A wireless LAN can be quickly created, but hard-wired networks are still a faster choice.

Networking multiple DAWs enables a new level of creative power: One computer can be used as a hard-disk multitrack recorder while the second is simultaneously running several software synthesis instruments. And new technologies are available (that we talk about shortly) to allow the horsepower of multiple computers to be pooled so that they can function as a single DAW.

mLan. Yamaha's mLan technology represents an extension to the FireWire standard that enables high-bandwidth audio and control connections between all kinds of studio equipment. The impetus behind its development is the replacement of the many different kinds of cables found in a studio with a single kind of cable for simplicity.

High-end solutions. For those with more money to spend, one of the ultra-fast, proprietary fiberoptic digital audio interfaces made by several companies may be a good choice. Or perhaps a custom infrared or satellite link to connect a DAW in one building to a DAW in another building is a better choice! The sky, so to speak, is the limit.

For Further Study

Bagwell C. "Audio File Formats FAQ." http://sox.sourceforge.net/AudioFormats.html.

Bosi M, Goldberg RE. *Introduction to Digital Audio Coding and Standards*. New York: Kluwer Academic Publishers; 2002.

Ebrahimi T, Pereira F. *The MPEG-4 Book*. Upper Saddle River, NJ: Prentice-Hall; 2002.

Franklin R. "Workstation File-Format Interchange, Part 1." *Mix* October 1, 2002.

Fries B. *The MP3 and Internet Audio Handbook*. Burtonsville, Maryland: TeamComBooks; 2000.

Hacker S. *MP3: The Definitive Guide*. Sebastopol, California: O'Reilly and Associates; 2000.

http://www.audiocoding.com

Pope ST, Van Rossum G. "Machine Tongues XVIII. A Child's Garden of Sound File Formats." *Computer Music Journal* 1995;19(1):25–63.

Exercises and Classroom Discussion

1. Convert a sound file into MP3 format at several different bit rates. Comment on any audible differences among the resulting files, and compare them to the original.
2. Calculate the residue, or error, between each MP3 file and the original audio file in Exercise #1. (Hint: you will need to think of a

way to subtract the sample values in one file from the sample values in another file.)

3. How many megabytes are required to store each of the following files?
   a. 5 minutes of stereo 16-bit, 44.1-kHz AIFF format data
   b. 5 minutes of stereo 24-bit, 192-kHz uncompressed WAV format data
   c. 5 minutes of stereo 192-kbit/sec MP3 format data
   d. 5 minutes of 8-channel 48-kHz AIFF format data

4. Perform the following tasks using royalty-free or noncopyrighted material:
   a. Rip a CD track, reverse it, and burn the result onto a new CD.
   b. Rip the soundtrack from a DVD-Video track, reverse it, and burn the result onto a new DVD-R.
   c. Rip the soundtrack from a DVD-Audio track, invert it, and burn the result onto a new DVD-R.

5. Perform the following tasks using royalty-free or noncopyrighted material:
   a. Rip a CD track as an AIFF or WAV file.
   b. Convert it to MP3, Ogg Vorbis, and AAC formats at identical bit rates.
   c. Construct an experiment to test others' preferences.
   d. Compare the results of running the experiment on a number of people.

6. Search the Internet to find as many freeware programs as you can for your computer that will allow you to convert an AIFF file into a WAV file and vice-versa.

7. What is the analog-audio-world equivalent of the digital-audio sound file?

8. Locate a free applications programming interface (API) for your operating system that will allow you to open, edit, and write AIFF or WAV sound files in your programming language of choice.

9. Suppose someone sent you an 8-channel 16-bit AIFF file. How could you split the sound file into eight separate mono sound files? Conversely, how could you construct an 8-channel AIFF file from eight independent mono sound files? Find or write a program to do this.

Editing and Mixing

In this chapter, we get to the "nuts and bolts" of using digital audio workstations (DAWs) for creative musical work. We first take a whirlwind tour of the basics of digital signal processing, which is essential for understanding the operation of the DAW and especially for creating additional tools for its toolbox. We then turn our attention to specifics of currently available hardware and software, including an exploration of the world of plug-ins. The chapter continues with a discussion of the DAW-based stereo project studio and the increasingly popular (and portable) laptop-based DAW studio. We conclude by discussing some basic mixing techniques and the importance of learning to mix through creative experimentation.

Introduction to Digital Audio Signal Processing

Crucial to the successful technical and creative use of the modern DAW is at least a basic understanding of the ways computers can process individual samples of digital audio, manipulating the numbers that represent each sample in various ways to create musical effects. *Digital signal processing*, or DSP for short, has been applied to many fields of study, including engineering, music, medicine, and economics. At its simplest, digital signal processing is nothing but the manipulation of sequences of numbers that represent meaningful, information-bearing data. Using the tools of DSP, we can do things like smooth out the kinks and rough spots in a signal, analyze it in various ways, reverse it, and mix it with other signals. The art and science of DSP forms the basis for the DAW, and the availability and power of DSP is the primary reason for working with audio in the digital domain.

Basics of Digital Filters

A filter is simply something that alters a signal in some way. A signal is a carrier of information, and signals can be conveyed in many ways: as electrical voltage or current, as a slew of numbers, as pits and grooves on a compact disc, or using anything else imaginable. When a signal assumes continuous values (that is, there are no gaps in values), we say it is an analog signal. When a signal assumes only discrete values (that is, gaps in possible values exist), we say it is a digital signal.

Digital filters therefore simply alter the information contained in a digital signal in some way. They can shape the sound of a musical signal by boosting some frequencies and cutting others. They can be used to create high-quality reverberation, delay, echo, and pitch-shifting effects, just to name a few examples. Filters come in many flavors: analog and digital, hardware and software. Analog filters are comprised of physical components like resistors, capacitors, and inductors. They work great in many applications and have played a remarkable role in music and audio over the years. However, we can now write filters in software, for instance in an audio plug-in, and we can even implement them on small hardware digital signal processing chips (like the one that is in your home theater receiver right now). DAWs, because they deal exclusively with digital audio by definition, use digital filters.

A digital filter applies a mathematical formula to an input signal to create a new output signal. The mathematical formula that represents the filter is applied to each input sample. The number of output samples need not be the same as the number of input samples. For example, a simple digital filter might discard every other input sample. The most basic kind of filters, and those which the vast majority of current digital audio workstation commercial software and plug-ins use, are called *linear, time-invariant* (LTI) filters. That is to say, the output is a linear combination of input terms, and the mathematical formula that represents the filter does not change over time.

A fundamental concept relevant to filters is the *transfer function*. A transfer function is nothing more than the ratio of the output of a filter to its input. Let's start with a simple example. If we feed the number 4 into a box, the box pops out the number 2. Next, we feed the number 6 into it, and the box pops out a 3. Clearly, the box is multiplying each number by $\frac{1}{2}$, and so we say that the transfer function of the box—the ratio of the output to the input—is $\frac{1}{2}$. (See Figure 4.1.) Said another way, the input to a filter, when multiplied by its transfer function, gives us the output.

Digital filters can be expressed in different ways; one way is as a transfer function as described above. Another way is as an algebraic expression; some simple algebra can help us characterize more complicated filters. In the algebraic representation, we denote the collective input signal to the filter by the letter x, and the output signal from the filter is usually denoted by the letter y. The first sample of x

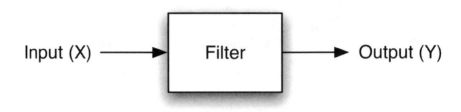

Input (X) ⟶ Filter ⟶ Output (Y)

is called x_1, the second sample is called x_2, and the nth sample is called x_n; ditto for y.

Here's a simple digital filter expressed in algebraic form:

$$y_t = x_t + x_{t-1}$$

In prose, this filter says, "The current output sample of the filter is computed by adding the current and previous input samples." How do would you suspect this filter might change the frequency content of an input signal? How might it change the amplitude of an input signal?

Let's rewrite this filter slightly by multiplying the right-hand side by $\frac{1}{2}$:

$$y_t = \frac{1}{2}(x_t + x_{t-1})$$

Now, the filter is simply averaging adjacent input samples, and so we call this simple digital filter a *two-point moving-average filter*. (We can construct moving-average filters of varying sizes by varying the number of previous samples to include in the average.) This filter says, "The current output sample of the filter is the arithmetic mean, or average, of the current and previous input samples." For the economically minded, this is the same process that Wall Street analysts use to calculate daily moving averages of financial data.

Consider the effect of this filter on a digital audio signal. For the lowest frequency possible (0 Hz, or DC), every adjacent sample value is identical; there is no variation in the signal. This means that the moving average of each pair is constant—it has the same as the value of each sample alone. That is, the filter has absolutely no effect on 0 Hz! The filter is said to pass DC, or allow 0 Hz to come through. For very low frequencies, the values of adjacent samples change very slowly, and so the filter's output is almost the same as its input.

With high frequencies, on the other hand, sample values change very rapidly. The highest frequency representable in digital form (which we already said was exactly one-half the sample rate, owing to the Nyquist Theorem) occurs when adjacent sample values alternate between the positive and negative extremes. (See Figure 4.2.) What is the two-point moving average of positive and negative versions of the same number? Precisely zero. This means that the two-point moving-average filter completely filters out the highest (i.e., Nyquist) frequency.

To summarize, the two-point moving-average filter passes DC unchanged in amplitude (i.e., with unity gain, that is, a 1:1 ratio), passes low frequencies fairly well, and eliminates the Nyquist frequency completely. We could continue our thought-experiment, and we would discover that the transfer function of the filter looks like a curve from 1 down to zero, as shown in Figure 4.3.

What would happen if we change the filter's expression to this?

$$y_t = \tfrac{1}{2}(x_t - x_{t-1})$$

Notice the "+" sign from before has been changed to a "−" sign. We now have a two-point moving-difference filter, rather than an averaging filter. The effect of this type of filter is left for you to discover in the problems at the end of the chapter.

Two Families of Filters

There exist many categories of digital filters. The most basic, as we mentioned earlier, are linear, time-invariant filters. Other filters, such as *adaptive filters*, can change the way they operate (i.e., change their own transfer functions) depending on their input. Many other broad families like these exist.

But we've only concerned ourselves with LTI filters here, which themselves are remarkably powerful tools for the modern DAW. Two basic categories of LTI filters are *feedforward filters* and *feedback filters*.

Feedforward filters. With feedforward filters, the output of the filter is a function exclusively of the input. Said another way, the present output sample is a function only of previous input samples. Feedforward filters are also called *finite-impulse response* (FIR) filters, because their output will eventually die down to zero given a

Figure 4.2
A two-point moving-average filter simply computes the average of each pair of input samples. The effect of a two-point moving-average filter on (a) 0 Hz and (b) the Nyquist frequency (half the sampling rate).

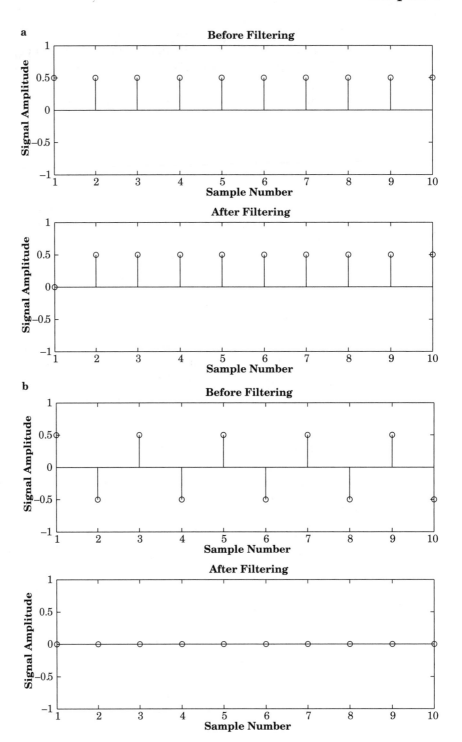

Figure 4.3
Simplified transfer
function of two-
point moving-
average filter. (a)
Numerical scale
plot; (b) decibel
scale plot.

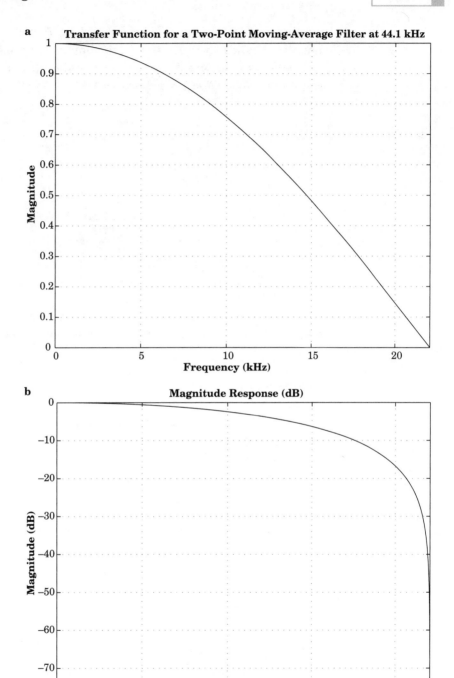

Figure 4.3
Simplified transfer function of two-point moving-average filter. (a) Numerical scale plot; (b) decibel scale plot.

single-sample "whack," or impulse. The filter's response to the impulse is called…you guessed it, the impulse response.

Feedforward filters are often represented graphically in a representation called a *flowgraph* (Figure 4.4).

Figure 4.4
Simple feedforward (FIR) flowgraph.

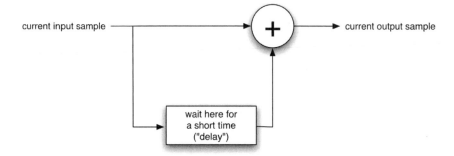

Feedforward filters form the basis for many musical effects, like equalization, convolution, and many others, which are discussed later.

Feedback filters. With feedback filters, the output of the filter is a function both of the input to the filter and previous outputs of the filter itself. This is shown graphically in Figure 4.5.

Figure 4.5
Simple feedback (IIR) filter flowgraph.

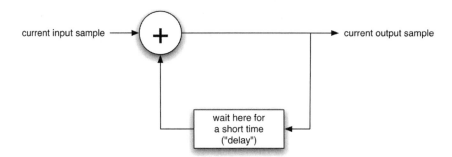

Feedback filters are also called *infinite-impulse response* (IIR) filters, because their output never completely dies down to zero owing to their inherent feedback. (Think of a delay effect in which each echo of a sound is reduced in amplitude by 50 percent. The amplitude would never exactly reach zero in a strict sense.)

IIR filters form the basis of many musical effects as well, for example delay effects, comb filtering, and artificial reverberation. Feedback and feedforward filters are often combined in various ways to create new effects, but the resulting filter is properly called a feedback filter, since its output is now (at least partially) based on the filter's own output as well.

We've only scratched the surface of the world of digital signal processing. If you are interested in learning more, there are some really great books on the subject, some of which are mentioned at the end of this chapter. Some of them even include extensive discussions of music and audio signal processing. Our rudimentary discussion of digital signal processing may seem a little esoteric at this point, but it will come in useful soon. A general understanding of digital signal processing will empower your understanding of the ways DAW software works on audio signals, and it will particularly provide insight into the creative use of DAW plug-ins, which we soon discuss.

DAW Software

A seemingly infinite number of software programs are available for modern computers to work with sound. Many are completely free, some cost a relatively small amount, and others may cost many thousands of dollars. Beware that price is not always a primary determinant of quality; some of the best DAW software available costs relatively little, and just because a program is expensive does not ensure its sonic quality. Price is of course often set by the intended audience for the product (academic, home-studio, or commercial) and any number of market conditions, including the number and availability of other similar products. With that said, however, a remarkable wealth of software resources exists for the modern DAW at all price points.

Software for the DAW tends to fall into five broad categories: two-track sound file editors, multitrack editors and sequencers, plug-ins, software sound synthesis environments, and miscellaneous programs that supplement the DAW in some way or defy categorization. We discuss two-track editors, multitrack editor-sequencers, and plug-ins in this chapter; software sound synthesis and miscellaneous software is left for Chapter 7.

Conceptual Overview and Terminology

As mentioned, all DAW software relies on the tricks and magic of digital signal processing to achieve its goals. And whether the software is intended to edit only stereo audio files (as with a simple two-track editor) or mix and process hundreds of audio tracks, many of the basic concepts are the same.

Destructive versus nondestructive editing. DAW software for editing and manipulating sound files falls into two broad categories: destructive and nondestructive editors. With each edit that a user makes, a destructive editor replaces the existing sound file on disk with the new, edited version. This overwrites the existing copy of the file, and an "undo" function is typically not available. The only way to undo an edit with a destructive sound file editor is for the editor to re-compute the original file's samples by computationally undoing the edit.

Nondestructive editors, on the other hand, keep the original sound file on the disk and simply write "patches," or changes to this file, as separate files. While this may take slightly longer than a simple destructive editor, the advantage is that nondestructive editors can allow an "undo" function—even multiple levels of undo.

Another advantage of nondestructive editing is that it allows *regions* of audio to be defined and then arranged and rearranged at will—all while leaving the original disk file alone.

Plug-ins. A plug-in is simply a software component that in some way enhances the functionality of a larger software program or environment. In the context of DAWs, the term plug-in refers to a software component that enables audio software to do stuff it was not originally programmed to do. This "stuff" could be emulating a tube distortion box, a reverberation processor, or mimicking a hardware audio level meter. The world of plug-ins is so vast that we devote the entire next chapter to them.

Basic sound editing. Most modern DAW software incorporates a standard paradigm for selecting, copying, cutting, and pasting portions of a sound file. This paradigm mimics the standard way in which most word-processing programs work. Selecting a portion of sound to edit is generally as simple as "clicking and dragging." Most graphical editors, whether two-track or multitrack, employ some or all of the following functionality and use the same or similar terminology.

■ **Select.** The user moves the computer's pointing device (mouse, trackball, trackpad, etc.,) to the desired beginning point of the edit. Pressing the pointing device's primary button and dragging to the left or right will highlight the corresponding graphical display of audio samples. Releasing the pointing device's button will leave the selected area highlighted (Figure 4.6).

Figure 4.6

Selecting a region of samples in a sound file.

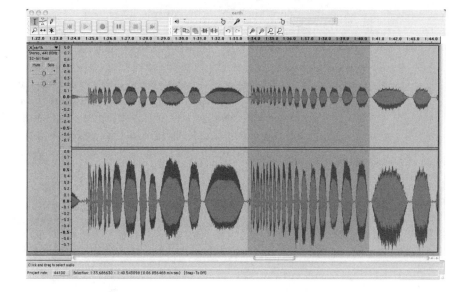

■ **Copy.** Once a region of audio samples has been selected, the user can copy it to the "clipboard," a kind of scratchpad memory storage space, for later recall. Audio on the clipboard can typically be pasted or inserted elsewhere in the file or in another file (or even another program altogether).

■ **Paste.** Most editors provide two ways to "paste" the contents of the clipboard into a sound file. In the first way, the user does not select a region of audio but simply clicks once at the desired insertion point. Invoking the paste function then inserts the contents of the clipboard at the desired location. In the second way, the user selects a region of audio; invoking the paste function then overwrites the selected region with the contents of the clipboard. Note that the selected and pasted audio need not contain the same number of samples.

■ **Delete.** Once a region of audio has been selected, it usually may be deleted by simply pressing the computer's delete key. Note that a click or pop may result if the deletion leads to a large discontinuity in sample values. In general, it is best to delete audio in this way:

1. Zoom in as far as possible so that the editor displays actual sample values.
2. Select the beginning of the region to be deleted at a zero-crossing (i.e., a location where the sample has a value of zero).
3. Zoom back out to the desired level.
4. Drag-select to the approximate ending location of the region to be deleted.
5. Zoom back into the sample level.
6. Fine-tune the ending location of the region to be deleted by drag-selecting to the nearest zero-crossing location.

This process is illustrated graphically in Figure 4.7. While it sounds tedious, following these steps helps ensure that the deletion process does not result in an audible click. Note that some editors include an optional "snap to the nearest zero crossings" feature that automatically performs the aforementioned steps.

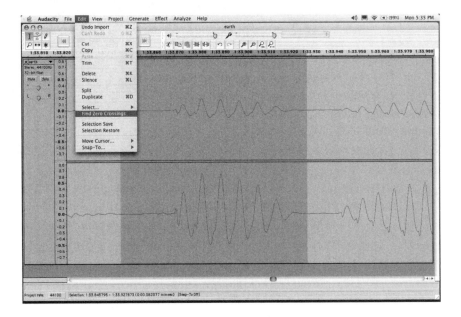

Figure 4.7
Zooming in to begin a cut at a zero crossing.

■ **Silence.** Many sound-editing programs include a silence function whereby all of the selected audio samples are silenced (i.e., their sample values are all set to zero). This can be useful for eliminating extraneous background noise from a sound file without changing its overall length (see Figure 4.8).

Figure 4.8
Most two-track sound file editors feature a "silence" function that sets the amplitudes of all selected samples to zero.

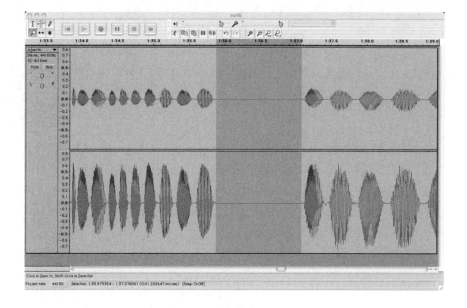

■ **Crop.** Just as one crops the edges out of a photograph, the crop function of many audio editing programs deletes everything in a sound file except the portion that is selected. This feature is useful, for example, when extracting a word from an entire sentence of recorded speech, or when selecting a great guitar riff out of a longer recording session. Be careful, though: If the sample values at the endpoints of the region to be cropped are not identical (or at least close enough in value), an audible pop or click can result, as we previously discussed.

■ **Audition.** Most audio editors allow the user to listen to a selected region of audio, usually simply by pressing a play button. If no region of audio is selected, pressing the program's play button (or often just the space bar) will begin audio playback from the current location.

■ **Scrubbing.** The term "scrubbing" has been carried over from the days of analog tape music in which precise cue points on the tape had to be found by manually dragging the tape forward and backward over the playback head in succession until the desired audio point for an edit was located. Scrubbing on most DAW software is as simple as dragging a "scrub tool" across the graphical waveform display to find a cue point. When scrubbing forward in time, you will hear the audio play back at a sample rate proportional to the speed at which your cursor moves across the waveform; when scrubbing backward, the audio similarly plays back at a proportional rate, but time-reversed.

- **Punching.** When recording a performance into a DAW for later editing within a mix, it frequently occurs that only a small portion of the recording is not usable. Most of the take may have been perfect and difficult to recreate, but a small portion may have been deemed unusable owing to simple performance error (for example, the wrong note was played) or sudden extraneous background noise (for example, a cough). With punching, only the offending portion need be re-recorded. The DAW can play back the track, and the performer simply "punches in" and begins playing at the desired time. The performer then "punches out" at the conclusion of the desired material. Many two-track editors and most multi-track editor-sequencers allow punch editing.

  Punching can either be automated or invoked with a trigger in real time. In automated mode, the mix engineer tells the DAW software the precise times of the punch in and punch out edit points. In trigger mode, the mix engineer or performer punches in and out with a physical trigger, such as a key on the computer keyboard, a footswitch, or simply by beginning to play.

- **Overdubbing.** Just as in the analog world, most DAW multitrack editor-sequencers allow a recording to be dubbed over an existing recording. Overdubbing allows layers of sound to be built from a single source. With overdubbing, one singer can be made into an entire chorus and a single violin can be made into the string section of an orchestra.

- **Looping.** Looping allows repetitive structures (for example, grooves for drum tracks) to be built more or less automatically from one short snippet or phrase of audio. In a typical use, a recording engineer might record a drummer playing several grooves. The mix engineer might then select a particularly good passage and loop it to create a longer passage. Looping a longer rather than shorter section can often mask the perception that the track was created from a loop, if that is the intent of the mix engineer. (Some drum tracks in popular music, however, often creatively exploit the static nature that a loop can create with great success.) When creating a loop, it is customary to ensure that the beginning and ending points of the loop have the same numerical sample value (often at a zero crossing) to avoid any clicks or pops. Looping a selection in which the surrounding audio is also of similar overall amplitude can also prevent undesired amplitude fluctuations in the resulting loop as well.

Fades. Even the most basic audio-editing programs generally include fade-in and fade-out functions, which do exactly as the name implies. Once a region of audio has been selected, applying a fade-in will modulate the amplitude (i.e., multiply the sample values) of the region by a continuous ramp function from zero to unity. Different ramp functions may usually be invoked, for example linear, exponential, or logarithmic, and many programs allow users to define their own ramp functions. The fade-out function does exactly the opposite: It scales the selected audio by a ramp function from unity to zero.

Logarithmic fades tend to sound the most natural to our ears, because the sound intensity level that we perceive is a logarithmic function of a signal's amplitude. Said another way, our ears respond to variations in acoustic power, not amplitude. (We will talk a little more about this in the next chapter.)

It is usually a good idea to apply a fade-in at the beginning and a fade-out at the end (at least of a few milliseconds) of any audio tracks. This ensures that playback of that track will not cause a click or pop when playing it back due to a discontinuity of sample values at the beginning or end of the track.

Multitrack editing programs for the DAW, and some two-track editors as well, provide the ability to cross-fade sound files. Cross-fading is often employed when two sound files are placed back-to-back with a slight overlapping region; a cross-fade encapsulates a fade-out of the first track while fading in the second track. It takes a bit of practice to learn how and where to employ them, however. With a good cross-fade, one can seamlessly splice two recordings or takes together.

A simple digital filter is used to fade a sound file in or out:

$$y_t = R_t x_t$$

where R_t is a ramp function that represents the shape of the fade. If R_t is linear, we have a linear fade; if it is logarithmic, we have a logarithmic fade. For example, if we define R_t to be

$$Rt = \frac{44,100}{t}, \ 0 \leqslant t \leqslant 44,100$$

we have a linear fade-in from 0 (no signal) to 1 (full value) over 44,100 samples. If the sampling rate is 44.1 kHz, this fade-in will take one full second.

Cross-fading between two sounds x and x' is simple as well:

$$y_t = R_t x_t + (1 - R_t) x'_t$$

Will you ever encounter filters like this? Probably not explicitly in this form, unless you are a programmer or want to write your own DAW plug-ins. In fact, you will encounter equations like this all over the place if you write plug-ins for DAWs. But many programs now allow the user to specify elements of filters in a form similar to this. For example, many two-track sound file editors allow user-definable fade shapes; that is, they let the user specify the shape of R_t in the above equations.

DC offset removal. As we mentioned earlier in the introduction to digital signal processing, the lowest frequency possible (0 Hz) is also referred to as DC (for "direct current," a term borrowed from electrical engineering). DC carries no information, either musical or otherwise; it is not a signal, because it does not contain any time-varying values. Compare this to a 440-Hz sine wave in Figure 4.9.

Figure 4.9
Time-domain
audio signal
representations.
(a) 440 Hz sine
wave;

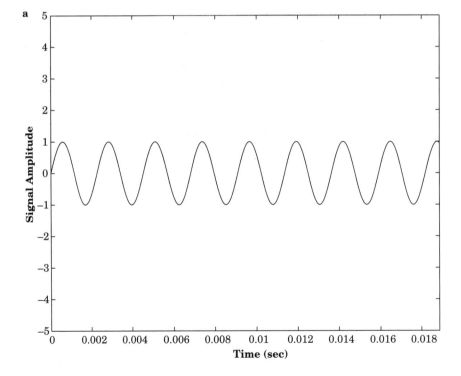

Time-domain
audio signal
representations.
(b) DC (0 Hz); (c)
440 Hz with DC
offset.

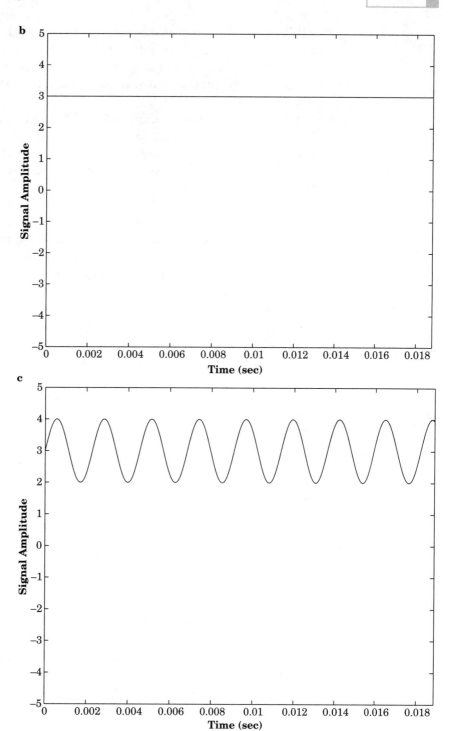

What would happen if you sent a DC voltage directly through an amplifier to a loudspeaker? It would cause the loudspeaker's cone to push out or pull in (depending on if you sent a positive DC voltage or a negative DC voltage) and stay there. (Don't try this on any expensive speakers!)

In general, the time-average value of the audio samples of a signal should be zero. This makes sense, because ideal audio signals contain ideal sinusoids of varying amplitudes and phases; the time of average of each of these individual sinusoids is zero, and so the average of their sum should be zero.

However, sometimes DC creeps into a sound file, often through an improper recording technique or by selecting a short portion of sound whose average sample value is nonzero. Invoking the DC Offset Removal function of most audio-editing programs removes DC from the sound file by performing the following steps:

1. Calculate the average sample value by adding all sample values from the file and dividing by the total number of samples in the file.
2. Subtract this average from each sample in the sound file.

At the conclusion of these steps, the time-average of the sample values in the sound file will be zero, and so the DC offset will have been removed.

Signal inversion. Inverting a selection simply multiplies each sample value by −1, thereby flipping the graphical representation of the sound file vertically. This is usually only performed when there exist phase problems between tracks that are being played simultaneously. Here is a hypothetical, simplified scenario of the relative phase of two signals contributing to a potential mixing problem (Figure 4.10).

Both signals in Figure 4.10 are sinusoids of frequency 440 Hz; they are out of phase by 180°. When the signals are played simultaneously, we hear nothing. Because mixing simply adds sample values of signals together, the resulting mix is a bunch of zeroes. If this is not the desired result, inverting one of the signals ensures that they add together so that we actually hear them.

Of course, we usually don't just mix sine waves. But occasionally corresponding frequencies of different sound files will be somewhat out of phase with each other such that a particular frequency is some-

Figure 4.10
(a) 440 Hz; 440
Hz 180° out of
phase with
respect to (a).

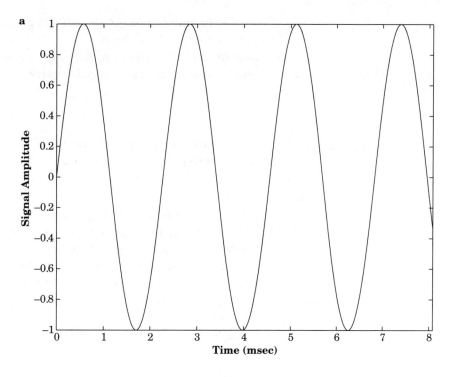

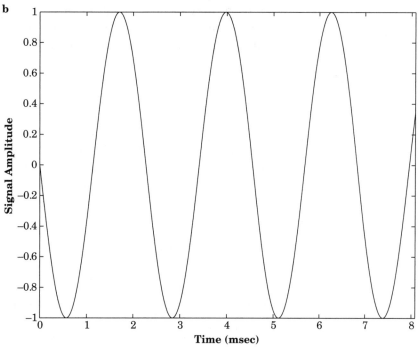

what quieter in the mix than in each individual sound file. Inverting one of the signals often fixes the problem. (It is a good idea for this reason to ensure your mix sounds good in mono as well as in stereo.)

The digital filter that inverts a signal is exceedingly simple:

$$y_t = -x_t$$

That is, the output sample at time t is simply the negative (inverse) of the corresponding input sample.

Normalization. It is frequently desirable to ensure that many sound files exhibit the same peak amplitude. In this case, we can normalize the files to the same level. The process of normalization applies the appropriate amount of gain or attenuation to a sound file such that it peaks at precisely the specified level. The normalization level can usually be specified as a percentage of the highest possible amplitude (which is of course dictated by the bit depth of the sound file), as a decibel level relative to full amplitude, or as a numeric sample value.

Normalization is also frequently used to raise or lower the apparent loudness of a sound file. However, because the process raises or lowers all sample values, it can affect the *noise floor*, or underlying noise level, present in the signal. Boosting all sample values raises the noise floor, and attenuating all sample values lowers the noise floor.[1]

When normalizing a sound file, the following steps are performed automatically:

1. Scan each sample in the sound file and find the one with the largest value.
2. Find the required gain or attenuation by computing the ratio of the requested numeric normalization level to the peak sample value.
3. Multiply each sample value in the sound file by the computed ratio.

Visualizing Sound

With the exception of entirely text-based DAW software, most programs for working with sound on a computer include some sort of

---

[1] This is true of integer sound files only. Using floating-point files avoids this problem.

graphical display of sonic data. These programs often vary greatly in the kinds of data they display and the way in which they display it. This process of finding ways to creatively and effectively visualize auditory data is precisely the reverse of sonification, which is the use of audio data (other than speech) to represent data. In fact, it has been shown that we are adept at hearing certain patterns, for example in stock market indices, that we cannot see by looking at visual representations on paper.

In general, displays of audio data on a computer screen take one of four forms: time-domain displays, frequency-domain displays, combined time-frequency domain displays, and experimental displays.

Most DAW software programs include facilities for viewing the basic time-domain and frequency-domain components of sound files, and some include time-frequency displays. Other types of displays are available as audio plug-ins for popular sound editors and mixing programs, and these are discussed later in the chapter when we turn our discussion exclusively to the world of plug-ins.

One concept central to the effective use of auditory displays on the DAW is related to terminology borrowed from the traditional recording studio console. The channel-level meters found on most hardware audio production consoles can be invoked in *pre-fader mode* or *post-fader mode*. In pre-fader mode, the level meter displays the audio signal level entering the corresponding channel. In post-fader mode, the level meter displays the audio signal level after the fader has been applied to it—that is, after the signal's level has been amplified or attenuated by the channel fader. This same terminology is used in auditory displays for DAWs, in that many displays can be invoked in either of these modes.

Time-domain displays. Time-domain displays of sound file data are probably the most commonly encountered form of auditory display in sound file editors and multitrack mixing programs. They are computationally efficient to draw onscreen and provide a clear, efficient visual reference for editing sounds.

These displays generally form the main graphical window of two-track editors, an example of which is shown in Figure 4.11. The numerical value of each sample present in an audio file is simply graphed on a two-dimensional plane. The horizontal (x) axis represents the passage of time, and the vertical (y) axis represents the numerical sample value. Notice that the vertical axis displays both positive and negative sample values, which correspond directly to the measured compressions and rarefactions of air molecules represented in the recorded or synthesized sound file. Zooming vertically alters the numerical sample value (y-axis) display scale, and zooming horizontally alters the time value (x-axis) display.

Time-domain displays can offer great insight into structural elements of the sound file, such as attack transients, peak amplitude levels, and durations of steady-state portions. Many programs display the peak sample value and duration of any samples that are selected, making this task much easier and more accurate.

Figure 4.11
A screen shot of bias peak. From this representation, we can visually discern where the transients are, what the peak amplitude level is, and the duration of the steady-state (sustain) portion of the sound. Note the entire sound file is displayed in the top portion, and the zoomed selection is displayed in the larger, lower potion.

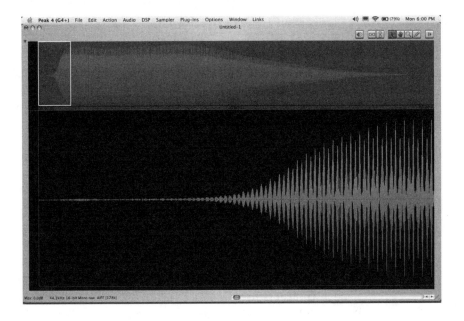

Frequency-domain displays. Viewing a sound file's data in the frequency domain can provide many insights into its structure that viewing it in the time domain cannot. As the name implies, frequency-domain representations of sound files show the spectral composition of the sound; that is, what frequencies are present in the sound and how "strong" each frequency is. (Some frequency-domain displays also indicate the relative phases, or time-alignments, of each frequency as well.)

Frequency-domain displays generally display frequencies on one axis and spectral strength on the other axis. Consider the example display in Figure 4.12. Without even listening to the sound, we can tell the sound is very pitched and sinusoidal; it is said to be comprised of primarily tonal (as opposed to "noisy") components. By examining the x-axis values where the peaks occur, it is clear that the fundamental frequency is 468 Hz.

Figure 4.12
Frequency-domain display showing the level of discrete frequencies present in the sound file.

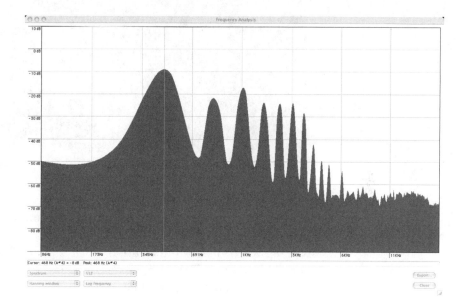

By contrast, consider Figure 4.13. We can immediately tell this sound is much more noise-like. Notice that all frequencies are present with roughly equal strength.

Most DAW software that includes built-in facilities for showing frequency-domain representations of sound files allows the spectrum of any selected portion of the sound file to be displayed, not just the entire sound file. However, the duration of the selected audio affects the accuracy of the display. Consider, for example, a sound file that contains a very low frequency, say 40 Hz. If the user tells the program to display the frequency spectrum of a 10-millisecond selection of a sound file, the frequency-domain display will not indicate any 40-Hz component in the signal. This is because a full cycle of a 40-Hz sinusoid cannot be completed in only 10 milliseconds. We say that the period, which is the reciprocal of frequency, does not fit within the selected window. The wavelength of a 40-Hz sinusoid is 1/40 =0.025 seconds, or 25 milliseconds.

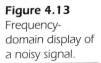

Figure 4.13
Frequency-
domain display of
a noisy signal.

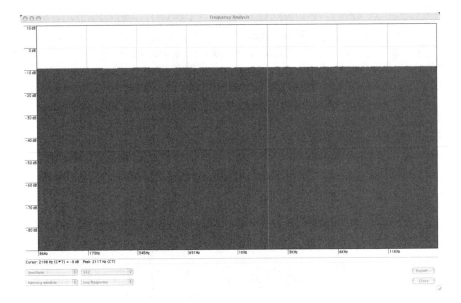

In general, the more audio we select, the more accurate the frequency-domain display will be. But this occurs at the expense of a smearing the transient information in the signal. This time-frequency tradeoff shows up all the time when working with sound on computers, and we talk about it more later.

Time-frequency displays. Combined time- and frequency-domain displays can be very useful when viewing a sound file. Most often, they are simply an animated, time-varying frequency-domain display. For example, a two-track DAW editor could compute the spectrum for each 0.25-second sliding window of audio samples in a sound file and display the data as the user navigates through the file.

Another common type of time-frequency display is called the *sonogram*, also called the *spectrogram*, an example of which we saw in Figure 1.5. Sonograms show frequencies along one axis (usually the vertical axis) and the passage of time along the other (usually the horizontal axis). The *spectral energy* (or *power spectral density*) present in each frequency component can then be displayed in one of two ways: (1) as a shade of gray in a grayscale sonogram; or (2) as a color hue, saturation, or intensity in a color sonogram. Sonograms may be computed out of real time and displayed as fixed, two-dimensional images, or they can be computed in real time and displayed as continuously moving images.

Some software, such as Audiosculpt from the Institut de Recherche et Coordination Acoustique/Musique (IRCAM, http://www.ircam.fr) and Melodyne from Celemony Software (http://www.celemony.com) allow users to edit sound files based on sonogram or sonogram-like representations. This can indeed be very useful when, for example, attempting to crop a portion of a sound file that exhibits a particular spectral shape, or when it is necessary to change the frequencies of the harmonics of a note irrespective of its fundamental.

Another time-frequency display is the *waterfall plot*. Waterfall plots are much like sonograms, except they indicate spectral strength as surface height along a three-dimensional surface. One axis represents the passage of time, another axis indicates frequency, and the third axis indicates the relative strength of the frequency components.

Two-Track Editors

Many mixing sessions begin by editing individual sounds or tracks of audio to remove excess noise, compress them, or normalize them, for example. When editing individual one- and two-channel sound files, a two-track editor built specifically for this purpose is most often used. As mentioned, most two-track editors follow the word-processing paradigm for editing sound files, making them quick, efficient, and easy to use. If you are comfortable with a modern word processor, it should be very straightforward to master a two-track sound editor fairly quickly.

Many DAW users tend to work in a more or less linear fashion on their sonic material. After a mono or stereo sound or track has either been recorded (either onto analog or digital tape, flash memory recorder, or directly into the computer) or synthesized directly inside the computer itself, it is then usually edited in a two-track editor to remove extraneous noise, any DC offsets, and the like. The file is then imported into a multitrack editor-sequencer for mixing. This "bottom-up" working paradigm is illustrated in Figure 4.14.

Furthermore, some two-track editors include the ability to batch edit sound files. With batch editing, the same edits can be applied in tandem to a group of sound files without individually opening and editing each one. For example, a program's batch editor might be instructed to change the sample rate and normalize to full amplitude every sound file contained within a folder. This can be enormously useful and save a great amount of time when working with large numbers of files. Two-track editors can also be useful in other tasks, like ripping CD audio tracks to disk.

Figure 4.14
The "bottom-up"
editing-mixing
paradigm.

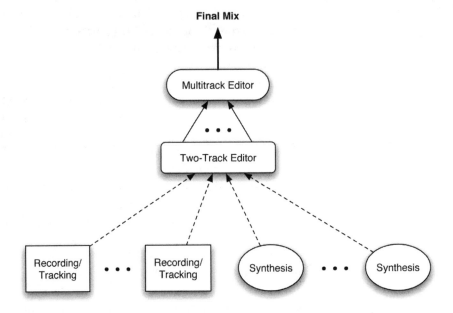

Figure 4.14
The "bottom-up" editing-mixing paradigm.

Two-track editors are also frequently referred to as mastering editors, because they are frequently employed when mastering a final two-channel mixdown of a multitrack session. Mastering is the final chain in the mixing process in which final mixes are collated, edited, and assembled into a single production for the target delivery medium—be it a physical disc or a sound file intended for streaming or download online. In most stereo mastering sessions, multitrack editors are not required, and so a two-track editor can be very useful.

We should now pause a moment to discuss commonly used two-track editing programs for different computer operating systems. Keep in mind that many multitrack editor-sequencers, which we discuss shortly, can also double as two-track editors if your budget is limited. The software listed here is intended only as a guide; not every available program is listed of course, and new software appears on the market frequently. In addition, software currently listed exclusively under one platform may soon be available for another; the industry is constantly expanding and in perpetual flux.

Macintosh-specific.   The standard two-track editor for the Macintosh is arguably Peak, from BIAS, Inc. (http://www.bias-inc.com). Peak can import and export virtually any file format one is likely to use, and its built-in ability to export AAC and MP3 files make it particularly useful when preparing audio for Internet broadcast and distribution. It allows an unlimited level of undo/redo, and it supports multiple-processor computers (which can be particularly helpful when applying reverberation and other processor-intensive effects). An entry-level version of the program, Peak LE, is also available at a reduced cost. Peak is also available in a plug-in version for multitrack editor/sequencers so that it can easily be used from within those programs.

Another very popular program in the Macintosh audio community is Spark from TC Works (http://www.tcelectronic.com). Spark includes many built-in features and effects that can be particularly useful when mastering, and it also supports batch processing of sound files. Of particular note is a built-in graphical effects routing matrix, called FX Machine, which allows as many effects as your computer will support to be chained together in any order.

DSP-Quattro (http://www.dsp-quattro.com), by Stefano Daino, is an inexpensive alternative to Peak and Spark for the Macintosh that also supports unlimited undo/redo and many real-time effects, including automatic tempo detection and automatic looping facilities.

A wealth of shareware and freeware editors is also available. Amadeus for MacOS X, by Martin Hairer (http://www.hairersoft.com), supports MP3 and Ogg Vorbis compressed audio formats. It also supports VST plug-ins, and it includes sound restoration and de-noising algorithms, support for multiple languages, and various sound analysis and synthesis functions. Alan Glenn's SndSampler (http://www.provide.net/~moorepower/ahg/sndsampler/) also supports a large number of sound file formats and includes batch sound-file editing capabilities. Sound Studio, by Felt Tip Software (http://www.felttip.com) supports 24-bit, 96-kHz audio and includes a wealth of effects plug-ins. It also supports a variety of sound file formats, including MP3 and AAC.

Some two-track editors are optimized for dealing with loops and grooves. Such programs, like ReCycle from Propellerhead Software (http://www.propellerheads.se), are feature-rich with tempo detectors, time- and pitch-shifting algorithms, and filters. Such programs can save composers and sound designers enormous amounts of time by automating many aspects of loop/groove composition and editing.

Windows-specific. The industry mainstay for Windows computers is perhaps Sony's SoundForge (http://mediasoftware.sonypictures.com), which has been on the market for over a decade. SoundForge supports unlimited undo/redo, high-definition audio (up to 64-bit floating point samples at 192 kHz), Direct-X audio plug-ins, and it includes a wide array of built-in real-time audio effects. Many users may find the vinyl restoration and built-in audio mastering tools particularly useful. The program also includes several spectrum analysis tools.

Adobe Audition (http://www.adobe.com), formerly known as Cool Edit Pro, is a slightly less expensive two-track editor for Windows computers that also works with high-definition audio. It too features many built-in real-time audio effects, and it integrates particularly well with other software from Adobe, like Premiere and After Effects. Sample-accurate editing, spectrum analysis tools, restoration and noise-reduction tools, and a built-in multichannel surround encoder make this program a compelling alternative to SoundForge. Audition is also a full-featured multi-track editor, although it is just as well suited for two-track editing.

WaveLab, a program from Steinberg Media Technologies (http://www.steinberg.de), is another popular Windows-specific sound file editor. Like other professional audio editing tools, it supports high-definition audio files and ships with its own suite of effects-processing algorithms. WaveLab includes an "Audio Montage" window, a nondestructive editing tool that enables users to quickly segment and re-order selections from an audio file. The program also includes real-time audio analysis tools to monitor input and output.

A cost-effective alternative to SoundForge, Audition, and WaveLab is GoldWave (http://www.goldwave.com), which is feature-rich with oscilloscopes, various effects, and a simple (albeit somewhat klunky) interface. It supports batch file processing and includes a noise-reduction algorithm. Particularly interesting is the "Expression Evaluator," which can synthesize a sound from mathematical equations supplied by the user.

Popular, free, two-track editors includes Acoustica from Acon Digital Media (http://www.aconas.com). Complete with plug-in support, signal analysis functions, audio restoration tools, and unlimited undo/redo, Acoustica could compete head-on with many high-priced commercial programs.

Should you use commercial software or freeware/shareware? We could debate the merits of each elsewhere, but suffice it to say that each has its own merits. In the past, I've devoted years to mastering a piece of shareware that no longer exists. The same can happen with commercial software, but the user guides and technical support of course tend to be much better. However, there is still something to be said for the tightly knit user communities that are most often formed around freeware/shareware, and because they are often open-source, the user community will often keep the project alive if the original author no longer works on the project.

Try software before spending money on it. Free demonstration versions are usually available for commercial software; compare them head-on with free software, and see which program suits you best. With so many great tools available, the choice of audio editing software is as much a matter of personal preference as anything else.

Linux-specific   Linux, the free UNIX-like operating system for PCs, has attracted a large community of very talented audio developers since its introduction in 1991, by then-computer-science student Linus Torvalds. Several audio-customized distributions of the operating system are now available, including one offered by the Center for Computer Research in Music and Acoustics (CCRMA; http://ccrma.stanford.edu) at Stanford University called PlannetCCRMA. Another, called AGNULA (for "A GNU[2]/Linux Audio Distribution"), is available at http://www.agnula.org.

A large number of free sound file editors are easily downloadable from the Internet. Some of these include AudioCutter Cinema from Virtual Worlds Productions (http://www.virtualworlds.de), GLAME (http://glame.sourceforge.net), and GNUSound (http://awacs.dhs.org/software/gnusound). Like much Linux freeware, GLAME and GNU-Sound have been collectively programmed and are maintained by a team of programmers over the Internet.

A comprehensive guide to Linux-specific sound-file editors, and all Linux audio software in general, is maintained by Dave Phillips online at http://linux-sound.org. The sheer number of interesting audio tools for Linux, coupled with the fact that the operating system is free, make Linux a very compelling choice for editing sound.

Cross-platform.   Recently, a group of programmers from around the world began working on a free, open-source, cross-platform two-track editor (Figure 4.15). Dubbed Audacity (http://audacity.sourceforge.net), the program includes much of the functionality of similar commercial programs; it supports WAV, AIFF, MP3, and Ogg-Vorbis sound files, unlimited undo/redo, and VST plug-ins, and it includes built-in audio effects. Furthermore, the entire source code is opensource, which may help ensure its long-term survival.

---

[2] "GNU" is the now-classic self-referencing acronym for "GNU's not UNIX," a class of open-source software, operating systems, and distribution licenses geared toward free, UNIX-like projects.

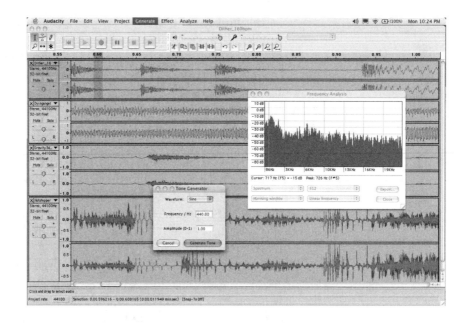

A cross-platform sound file editor that has been around for some time is WaveSurfer (http://www.speech.kth.se/wavesurfer), formerly known as *xs*. Developed by Kåre Sjölander and Jonas Beskow, speech researchers at the Royal Institute of Technology in Stockholm, WaveSurfer is a sound-file editor and analyzer particularly well-suited for working with speech. It supports a large number of audio file formats, and it supports additional file formats through plug-ins.

Multitrack Editor-Sequencers

In the previous section we mentioned one way of creating a mix from the bottom up by working first with individual sounds in a two-track editor and then importing them into a multitrack environment for mixing. When recording multiple sound sources simultaneously (such as an entire rock band or chamber ensemble) or when recording or synthesizing multichannel/surround audio files, a two-track editor is clearly not suitable: In these cases, we need to simultaneously work with more than two tracks of audio. In this case, the sounds are directly recorded and/or edited in a multitrack editor-sequencer. Fortunately, these programs generally offer the same built-in editing features found in two-track editors (e.g., DC-offset removal, signal inversion, normalization).

Whether or not a two-track editor is used in the recording/editing chain, a multitrack editor-sequencer is required to edit and sculpt the mix. These programs typically incorporate the following basic functionality:

1. Real-time mixing and playback of a large number of sound files
2. Real-time calculation and execution of fades and amplitude changes on each track
3. Automation of some or all mix parameters (like the amplitude of each track as a function of time, or the feedback amount in a delay plug-in as function of time)
4. Support for external control surfaces
5. A window that displays a graphical overview of the time-domain waveforms of each track
6. A window that displays a simulated mixing console graphically
7. At least basic recording, editing, and sequencing of Musical Instrument Digital Interface (MIDI) data
8. Audio effects editing and processing via both built-in effects and support for one or more software plug-in architectures

Some programs even incorporate their own built-in two-track editors as well, although these are rarely as robust as standalone two-track editing programs. Many also support the synchronized playback of video sequences, allowing one to sculpt a mix to correspond temporally to events in the video.

We use the term "editor-sequencer" to describe these programs, because not only can they edit sound files directly, they typically support MIDI sequencing as well. (We'll talk more about MIDI and other communications protocols in Chapter 9.) Some even include music notation facilities as a standard feature.

Mixing revisited.   Before we discuss specific programs, we should pause now to revisit the concept of mixing. In Chapter 1, we asked what "mixing sound" really means. With an understanding of binary numbers, sound files, and some of the basics of digital audio signal processing, we can now examine some of the technical aspects of digital audio mixing.

As a craft, mixing is a skill that takes a great amount of time and practice to develop. But from a purely technical perspective, however, digital audio mixing, at the fundamental level, simply involves adding together sample values from various sound files, applying digital filters (e.g., equalization, effects, and gain changes), and rescaling the sums of sample values so they do not "clip" or exceed their maximum word-length-allowable value.

Consider two mono sound files. When these two files are played back in a multitrack editor-sequencer, their sample values are simply added together. Different programs perform the sample addition in slightly different ways, but all programs must hold the summed sample values in a higher word-length array of values than the word length of each sound file. This is done to prevent *overflow*. For example, the highest sample value that a 16-bit unsigned, integer sound file can contain is 2^{16}, or 32,768. Clearly, the sum of two such numbers could potentially take several more than 16 bits to express. (For example, $2^{16} + 2^{16} = 32,768 + 32,768 = 65,536$, which takes 17 bits to represent.) Furthermore, the sum of many such numbers (as when many sound files are mixed together) could take many more bits to express. When a multitrack editor-sequencer advertises that it performs 32-bit processing, it means that that software stores the added sample values internally in 32-bit words. It also means that any effects processing is also performed internally with 32-bit words.

Once the sound files have been placed in the mix engineer's desired temporal alignment, the mix engineer might then proceed to adjust their relative amplitude levels with some sort of graphical curve in a multitrack editor-sequencer, as shown in Figure 4.16.

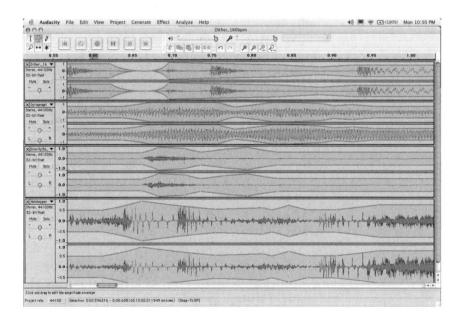

Figure 4.16

Adjusting the relative levels of two sound files in a multitrack editor-sequencer.

Perhaps the mix engineer then decides to distribute the sounds within the stereo or surround sound field and apply a plug-in to process each of the sound files. This process of time-aligning sound files, setting their relative amplitude levels, distributing them in the sound field, and applying plug-ins suggests a variety of possible orderings of the technical steps involved in creating a mix. No one process is inherently better than another; it's a matter of personal preference.

Working with a multitrack editor-sequencer. As with any program, the best way to learn your way around a multitrack editor-sequencer is to dive right in. Third-party educational products specific to each of the major programs are available, and many of them can be quite useful. For me, the best way to learn, though, has been through a combination of trial and error and apprenticing with an expert as much as possible. Simply watching an experienced "power user" mix with a multitrack editor-sequencer can be extremely instructive.

One of the more powerful features of multitrack editor-sequencers is that they allow the user customization of keyboard commands, and some allow user-definable macros to execute sequences of commands. For example, Digital Performer has the notion of "key sets" that allow users familiar with a different program to import the keyboard shortcuts from that program. This can make Digital Performer "feel" more like Pro Tools, for example, which is especially useful if one works in both environments frequently.

In general, you begin creating a mix in a multitrack editor-sequencer by creating a new mix file and importing sound files to include in the mix. These sound files are then usually just "dragged" to the desired tracks in a rough temporal alignment, and the fun—and work—then begins.

Another way that multitrack software is commonly used is in recording applications. When audio is recorded directly into multitrack software, the DAW is often referred to as a *hard-disk recording system*, because sounds are being recorded directly to the computer's hard disk (as opposed to, say, magnetic tape). Indeed, this is one of the most useful hats worn by the modern DAW. As such, we talk about hard-disk recording separately in a later chapter. For now, we just concentrate on the basic principles of working with multitrack software, assuming we already have sound files with which to work.

Working with multitrack software necessitates a basic understanding of the terminology it employs, much of which has been co-opted from traditional recording studio paradigms. Depending on your background, this section may be an insultingly simple review or a necessary bit of information.

TRACKS AND CHANNELS. One bit of terminology that can be the source of some confusion for DAW beginners is the distinction these programs make between *tracks* and *channels*. Similar terminology is used to distinguish *multitrack* from *multichannel* sound files. In general, a track refers to a sound file or sequence of sound files controlled by one unique mix fader. A multitrack editor-sequencer that claims to allow 128-track mixing means that 128 different sound files or sequences of sound files can be mixed simultaneously.

To complicate matters, the term *virtual tracks* is often used to denote tracks that "piggy-back" any extra DSP horsepower of the host system. For example, a multitrack editor-sequencer that allows 196 virtual tracks may allow 196 separate tracks of sound files, but perhaps only a subset of those can be mixed simultaneously. Many programs allow unlimited virtual tracks and attempt to play back as many tracks as the host computer or DSP accelerator card can.

The term *channels*, however, refers to the number of discrete output streams to which the final mix is ultimately reduced. For example, a two-channel mix means that all tracks have been mixed down to two channels. A 128-track mix may be reduced to a two-channel mix, for example. A stereo mix is a two-channel mix (even though it may be presented in concert or at home using more than two loudspeakers). A Dolby Digital mix is a 5.1-channel mix, meaning that it includes separate audio channels for front left, front center, front right, left surround, right surround, and subwoofer. (We return to this in detail in the next chapter, which specifically deals with surround mixing.)

DOWNMIXING AND UPMIXING. The process of mixing a number of tracks or channels down to a smaller number of channels is known as *downmixing*, and the process of mixing a number of channels into a greater number of tracks or channels is known as *upmixing*. An example of downmixing would be reducing a multichannel, surround-audio mix to a stereo mix. An example of upmixing would be producing a multichannel, surround-audio mix from a stereo mix. Both upmixing and downmixing are increasingly encountered in modern studios, particularly as composers, mix engineers, studios, and the marketplace embrace surround audio in the creation and delivery of music in various formats.

Interestingly, technology is available to automatically upmix and downmix, and several companies have been formed to address the related technical and artistic issues. Various signal-processing tricks are available to automate the process, with varying degrees of success, although this approach may not be the best for all musical situations. We also talk more about upmixing and downmixing in the next chapter on surround mixing.

PRIMARY USER INTERFACE WINDOWS. DAW multitrack editor/sequencers almost always feature at least two different fundamental user interface windows. Although different programs call them different names, one is usually called something like the "mix" window, and the other is usually called the something like the "edit" window. The mix window usually emulates the look and feel, albeit on a computer screen, of a more traditional hardware mixer. *Channel strips* (the thin vertical strips of faders, knobs, and buttons that control the audio on a particular track) are arranged in parallel, side-by-side, just as on a hardware console. From the mix window, the user can move faders, access the aux sends/returns and output busses, select plug-ins to apply to that track, and so on.

The edit window is an interface that graphically displays a time-domain representation of some or all of the tracks in the currently open mix. The user can typically zoom in or out, scroll left to right or top to bottom, drag and move sound files around, and many other things as well.

Many users frequently switch back and forth between these two graphical representations of a mix, so learning the appropriate keyboard shortcut in your multitrack software to activate each window can be a real timesaver.

SIGNAL FLOW. Just as with traditional hardware consoles, the signal flow—that is, the particular processing path through which audio flows—can vary among software multitrack mixers for the DAW. However, most multitrack software built around the traditional hardware console paradigm employs graphical representations of channel strips, as we just mentioned. And in general, the audio signal flow through each strip runs from top to bottom.

Figure 4.17a shows a block diagram from a typical channel strip; Figure 4.17b shows a screenshot from the channel strip of MOTU's Digital Performer. In general, the first (top-most) section provides *sends* and *returns*, which allow audio to be sent elsewhere for further

Figure 4.17
Channel strip. (a)
Block diagram; (b)
screenshot from
Digital Performer
channel strip.

a

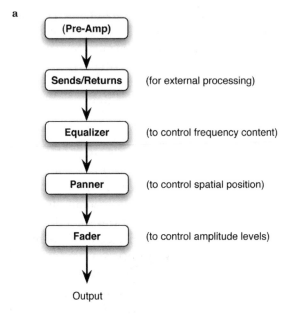

(Pre-Amp)

Sends/Returns   (for external processing)

Equalizer   (to control frequency content)

Panner   (to control spatial position)

Fader   (to control amplitude levels)

Output

b

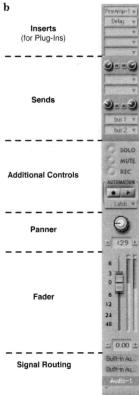

Inserts
(for Plug-Ins)

Sends

Additional Controls

Panner

Fader

Signal Routing

processing and output. Then, audio is processed by an *equalization* section to boost some frequencies while cutting others. Next, a *panpot* allows the mix engineer to control the spatial location of audio on that particular track. Finally, a *fader* allows control of the amplitude scale of audio on that track.

Note that any of these features on a channel strip can usually be automated, either by recording your movement of the virtual knobs, buttons, and faders in the mix window, or by graphically drawing curves with a pencil tool in the edit window.

Although the precise arrangement of the virtual channel strip (and the graphical mixing board that they comprise) varies among software programs, most of them tend to follow this same basic paradigm and general arrangement.

BUSSES, AUX SENDS/RETURNS, AND PLUG-INS. Unless you have worked in a traditional studio before entering the world of the DAW, you might not be familiar with the terms *bus*[3] and *aux send/return*. The notions of master tracks and plug-ins are generally pretty specific to the DAW, but they fit well into our discussion of busses and aux sends/returns.

In a traditional mixing console, a bus is simply a signal path for any number of tracks of audio. Said another way, busses allow audio to be "tapped" from one or more tracks and sent elsewhere for further processing. They also provide a mechanism allowing audio samples to be sent directly out the mixer (or in our case, the DAW's audio interface). A traditional four-output-bus mixer, for example, allows any of the channel strips to send audio to one of four of these paths, in addition to the *main output bus*, also called the *master outputs*. (Busses can be mono, stereo, or surround, but we'll leave out this complication for now.) With the DAW, the number of available output busses can be significantly higher, bounded only by the multitrack software you use, the speed of your computer, and the number of output audio jacks on your audio interface(s). Unless you need to multitrack record a large number of sound sources out to tape or play back 32-channel surround audio, however, most people get by fine with just a few output busses. Even the least expensive DAWs typically allow eight outputs. Output busses are typically named "Bus 1," "Bus 2," "Bus 3,"

---

[3] Borrowing from electrical engineering, "bus" is often spelled "buss," which is a conducting post that ties together multiple wires. In computer science, a buss is a signal path that connects components of the central processing unit (CPU) together. Interestingly enough, it was an Old English work that referred to kissing.

and so on, or they can be given customized names like "Front Left," "Surround Right," and so on.[4]

And here is where things get a little messy: There are two kinds of busses—*output busses* and *auxiliary ("aux") busses*. We mentioned that output busses allow audio to be sent out of the DAW, but so do aux busses. The only real difference is that aux busses allow a complementary signal return path for audio that is sent to them. This is why they are often called "aux sends/returns." Aux busses are usually used for packaging a group of tracks in a mix, processing them, and then returning the processed audio back to the group of tracks. Using controls provided by the aux busses and the audio processing mechanisms, we can control the relative amounts of "wet" and "dry" (processed and unprocessed) audio.

On the DAW, the general mechanisms by which we dynamically process audio samples are plug-ins. (We get to them in detail shortly, so hang on!) Plug-ins can be applied either directly to an individual audio track or to a bus. Applying a plug-in to an individual track invokes that plug-in's processing algorithm only on that particular track. On the other hand, applying a plug-in to an entire bus (whether an output bus or an aux bus) invokes that plug-in's algorithm on all tracks that comprise that particular bus. Assigning a plug-in to a bus rather than an individual track can save an enormous amount of computer processing power and is very advantageous in certain circumstances, but it has direct musical implications that we cover momentarily.

TRACKS REVISITED. I have skirted the issue a bit, but with a basic understanding of busses, aux sends/returns, and plug-ins, we need to re-examine the notion of tracks in multitrack mixers. Again, the specific implementations of the functionality discussed here varies somewhat from program to program, but the general concepts remain the same.

Recall that the term *channel* refers to a discrete audio output path of a finished mix. We can mix for 2 channels, 5.1 channels, or 16 channels, for example. A large number of tracks, however, may comprise the mix for each channel.

To incorporate the traditional mixing paradigms of tracks and busses with the DAW-specific technology of plug-ins and the Musical Instrument Digital Interface (MIDI) communications protocol, how-

---

[4] Most multitrack editors reserve labels like "Main Out," or "L" and "R," for the first two output busses, which correspond to the master outputs of a traditional console.

ever, multitrack mixers for the DAW typically allow the tracks that comprise a mix to be one of four basic types: *audio tracks*, *master tracks*, *auxiliary tracks*, and *MIDI tracks*. Note that the specific software you use may employ different names, but the concepts remain more or less the same.

Each track in a mix, no matter its type, is typically represented graphically in both the mix and edit windows. (Recall that the mix window displays a graphical representation of virtual channel strips and looks like a traditional mixer, while the edit window displays the time-domain audio samples of tracks in a top-to-bottom arrangement.)

Another feature of tracks, no matter their type, is that they allow plug-ins to be applied to their content. Audio plug-ins can be applied to audio tracks, aux tracks, and master tracks, while MIDI plug-ins can be applied to MIDI tracks.

Audio tracks most directly mimic the functionality of the individual channel strips on a traditional hardware mixer: They contain a single track of one or more audio files arranged linearly (left-to-right) in time. Each audio track in a mix can be represented graphically in the mix and edit windows. A major departure from traditional mixing consoles, however, is that audio tracks can contain not only monaural audio, but they can also be configured to contain stereo or even multichannel sound files. This is amazingly powerful, for with a single fader on a single track, we can now easily control the amplitude of a multichannel sound file that was recorded using surround recording techniques.

Different multitrack mixers support different audio file formats, and some allow multiple file formats; check the manual of your specific program for details on which formats are supported for audio tracks. In a simple pop-music mix, for example, one track might contain a sound file recording of the lead vocals, the next track might contain a recording of a bass guitar, another might contain lead guitar, and another the drum track. Note that only one sound file at a time may play on any single audio track; to play more than one sound file at a time, simply create new audio tracks and add sound files to them.

Multitrack mixers provide access to and control of the busses via two kinds of tracks: auxiliary tracks and master tracks. An auxiliary (aux) track allows control of a particular aux bus, while a master track allows control of a particular output bus. Again, plug-ins may be inserted into either of these kinds of tracks.

Aux tracks are typically used as a kind of virtual effects processing box. They are typically named "Aux 1," "Aux 2," "Aux 3," and so on, or they can be given customized names like "Reverb Processor" or "Distortion." For example, let's say that we create an aux track called "Delay" and apply an audio delay plug-in to it. Now, every audio track that is sent to the aux bus called "Delay" will be sent through an aux bus and processed in the same way by the delay plug-in. The processed audio samples will, in real time, be returned to and mixed with each sending track.

The aux send level—that is, the amplitude scale of the sending track's audio samples—can usually be controlled by either adjusting the sending track's audio fader or a dedicated aux send level control provided by the sending track. (The notion of pre-fader sends and post-fader sends becomes important here. We'll get back to this in a moment.) The aux return level is generally controlled by adjusting the fader position on the aux track. Manipulating these aux send/return levels allows dynamic control of the wet/dry level of a particular effect.

Just as aux tracks allow access to and control of the aux busses, master tracks do the same for the output busses. With a master track, any track assigned to that particular output bus can be globally controlled together. For example, if we are creating a simple stereo mixdown from 16 audio tracks, we should route each audio track to the master output bus and create a master track. Now, we can globally sculpt the amplitude range of the mix by controlling just the master track's fader. We can also globally equalize the mix by adjusting the equalizer only on the master track.

Earlier I mentioned that busses, whether output or aux, can be mono, stereo, or surround. And hence, both aux tracks and master tracks can be configured as mono, stereo, or surround. If we are processing a group of 5.1-channel surround-audio tracks with a multichannel reverb plug-in on a particular aux track, then we should configure that track as a 5.1-channel aux track, for example. Or if we are making an 8-channel mix, we might want to create an 8-channel master track to globally control the mix.

Note that multitrack mixers can vary greatly in their support of multichannel audio. Some do not support it at all, except as a hack, while some can easily accommodate up to 12.2-channel, or more, mixes. There are several tricks to work around this, which we address in Chapter 6.

Finally, multitrack editor-sequencers provide an integrated environment with which to sequence MIDI information and incorporate it within an audio mix. A MIDI track is simply a dedicated track that contains only MIDI information. MIDI tracks are often used to record or import MIDI sequences. If your DAW is connected to MIDI-aware equipment like synthesizers, then playing back your mix routes the audio-only portion of your mix through your DAW's audio interface while sending MIDI information out to MIDI equipment. When using MIDI tracks, a standalone hardware mixer can be useful in mixing together audio from the DAW with audio from MIDI output devices.

Of course, MIDI tracks can contain any kind of MIDI information and control a lot more than just synthesizers. Using a MIDI Time Code embedded on to a MIDI track, for instance, we can automatically trigger the "record" button on a multitrack recording machine when playing back a mix.

SUMMARY OF BUSSES AND TRACKS. The interplay of busses, tracks, and plug-ins may seem at first a little confusing. But a thorough understanding of how they work together can provide an unbounded number of creative possibilities when working with the DAW.

One sticky point is that most multitrack editors can be configured to allow user-customizable routing schemes between the busses and aux sends/returns in software and the physical hardware audio jacks on the DAW's audio interface. (Again, we talk more about this in the Chapter 6, as it becomes particularly important and useful when working with surround audio.) This flexibility can accommodate a wide variety of studio settings and personal working styles.

The general functions and uses of the different kinds of busses, tracks, and plug-ins are summarized in Table 4.1.

Let's now talk about some of the musical implications of the specific interplay of busses, tracks, and plug-ins. Two examples will help illustrate. Let's suppose Track 1 contains a recording of a singer that we would like to process with a reverberation plug-in, adjust the amplitude of the processed track, and then send it out the main output bus. In this example, we wouldn't need to create an aux track to process the singer; we would simply invoke a reverb plug-in on Track 1.

That was easy enough, but what if we have a 48-track mix, and we want to process each and every track with reverb? Clearly, invoking and configuring a reverb plug-in on each track would be tedious and time-consuming—not to mention incredibly taxing on our computer, because reverb algorithms are typically quite processor-intensive. We can make life easier by first creating an aux track, invoking and configuring the reverb plug-in only on the aux track, and then sending each track out to that aux track. Because each track is sent to the

TABLE 4.1   Overview of Busses, Aux Sends/Returns, and Plug-Ins

	Component	Function	Uses	Notes
Busses	Master Output Bus	The primary output bus in a multitrack mixer; allows all tracks assigned to the master output bus to be controlled by the master track	In stereo mixes, the master output bus is usually routed out the audio interface to head phones or monitor loudspeakers	
	Other Output Busses	Provides alternative output signal paths	Especially useful for surround-audio mixing, sound diffusion, and multichannel recording	Can also emulate the channel "direct out" jacks on a traditional mixer
	Auxiliary Busses ("Aux Sends/ Returns")	Allows a group of tracks to be controlled by another track (called an aux track)	Often used to process multiple tracks with the same plug-in	
Tracks	Audio Tracks	A holding tank for one or more sound files; only one sound file can play at a time; can process audio with the track's channel strip or plug-in(s)	The building blocks of a mix; this is where all the audio is placed	
	Auxiliary Tracks	Allows control of an aux bus; can thus control a group of audio tracks, optionally returning the processed audio to the sending tracks	Effects processing on a group of track simultaneously	
	Master Tracks	Allows control of an output bus		
	MIDI Tracks	A holding tank for MIDI information	Sending control information to MIDI equipment	
Plug-Ins ("Inserts")	Audio Plug-Ins	Allows an audio track, aux track, or master output track to be processed or examined	Effects processing, audio analysis and metering, etc.	
	MIDI Plug-Ins	Allows the MIDI data on a MIDI track to be processed or examined	MIDI effects processing	

same reverberation plug-in, the physical interpretation of this is that each track is placed in the same acoustical environment. On the other hand, if we wanted each track to be in a different acoustic space, we would need a separate reverb plug-in on each audio track, but this scenario is perhaps less often encountered.

In general, if a particular effect is only needed on one audio track, then it should be applied as a plug-in. If, however, the same effect is desired on multiple audio tracks, it should instead be used on an aux track. In addition, if the mix engineer wishes to apply the same effect to every track in a mix, then the plug-in should be inserted into the master track(s).

Another way of looking at the issue of when to use a plug-in on an individual track versus when to use it on an aux track involves the host computer's processing speed. Plug-ins that are relatively simple from a signal-processing complexity standpoint, such as delay, echo, and flanging, can usually be invoked on individual tracks—often a hundred or more different tracks at a time. On the other hand, more computationally intensive plug-ins like reverberation, as we mentioned, are often better suited for use on an aux track, unless your musical vision (and computer horsepower) allow otherwise.

PRE-FADER, POST-FADER. The terms *pre-fader* and *post-fader* are often encountered when working with multitrack software. In general, these terms apply to aux sends, plug-ins, and audio metering. (We address plug-ins and metering later.) Aux sends, plug-ins, and audio display meters can be typically be set to act in either pre-fader mode or post-fader mode.

Simply stated, pre-fader mode means that audio is sent out from a track (either to an aux send, a plug-in, or an audio display meter) independently of a track's amplitude fader. The raw audio samples are sent before they are attenuated or amplified by the corresponding fader. Post-fader mode means that audio is sent out from a track after it has been affected by that track's amplitude fader. (See Figure 4.18.)

Why is this important? Let's take a simple musical example. Consider an audio track that contains a recording of a vocal line; let's say we want to place the singer in a large acoustical space, like a big cathedral, by applying some reverberation to it. One final constraint for our example: We want to create a gentle fade-out of the vocal line.

For the reverberation effect, we could use a hardware reverberation processor or a comparable plug-in. Let's say our multitrack mixer program came with a great reverberation plug-in, so we will use that. Should we use pre-fader or post-fader mode for the plug-in?

Well, let's first create the fade-out on the vocal track, as shown in Figure 4.19. By drawing an amplitude curve (in this case a line) like this, we have simply automated the amplitude fader's position on that track. Now, we send the vocal track to our reverb plug-in.

Figure 4.18
(a) Aux send
operating in
pre-fader mode;
(b) aux send
operating in post-
fader mode.

a

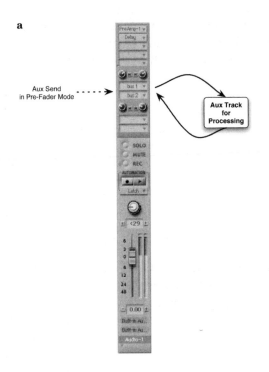

Aux Send
in Pre-Fader Mode

b

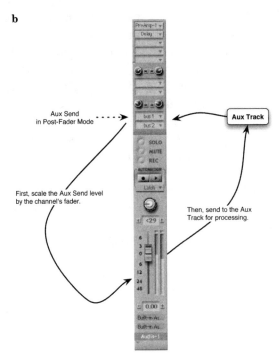

Aux Send
in Post-Fader Mode

First, scale the Aux Send level
by the channel's fader.

Then, send to the Aux
Track for processing.

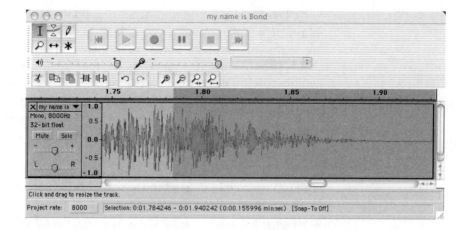

Figure 4.19
Fade-out of a
vocal line.

Let's consider the sonic differences of using pre- and post-fader in this example. If we sent the vocal track to the reverb in pre-fader mode, then the ratio of wet-to-dry (that is, the relative level of reverberant to unprocessed signal) in the sonic result will be constantly changing during the fade-out. In fact, because the audio is sent to the plug-in independently of the fade-out, we are in essence fading out the dry, unprocessed signal while maintaining the same level of wet, reverberant signal. What do you think this would sound like? To my ears, it could sound like the singer is moving farther away from me while the size of the cathedral is increasing. This could indeed be a very unique, if physically difficult or impossible, effect!

Using the plug-in in post-fader mode would clearly preserve the wet-to-dry ratio, because as the dry signal fades out, so does the amount of the dry signal that is being sent to the reverb plug-in. Hence, as the dry level fades out, so does the wet level. This would sound more like the singer is simply moving farther away from us within the same acoustic space.

AUTOMATION. Using the automation features of multitrack mixers, a user's specific actions to control virtually any element of a mix can be saved and recalled for later playback. This allows us to do things like record the level of each audio track as a function of time, or dynamically change the reverb time on a plug-in by moving a knob on a control surface (hopefully without causing clicks and pops, if the plug-in is designed well!), for example. Automation information can be recorded by remembering the exact actions the user made with the mouse (like moving a fader), or it can be recorded using a MIDI controller or external control surface, like a keyboard or mixing console controller.

At least three fundamental automation modes are typically found in multitrack editor/sequeners. Their specific names may vary according to the maker of the software, but they are often called *read mode*, *write mode*, *touch* (or *update*) *mode*, and *latch mode*, the same terms usually used in digital hardware consoles.

Read mode simply plays back any recorded automation. This is the mode used to audition the results of previously recorded automation. Without read mode invoked, then no recorded automation will be played back, or read.

The three common automation record modes have subtle differences. Write mode is used to record all information, even if a fader or knob is not moved. In write mode, the software begins recording all write-enabled parameters as soon as playback begins, overwriting all previously recorded automation.

With touch mode, information is recorded only when a fader or knob is "touched," either with a hardware knob or fader on a control surface or virtually with the computer mouse. Even if the fader is not moving, its position is being recorded as long as it is touched and held. Because touch mode records mix parameters only while a control is touched, the recording of automation information ceases as soon as the control or fader is released, and the mix parameter returns to its previously recorded location.

Latch mode is similar to touch mode, except that only changes in a control are recorded. Also, once the mix parameter has been changed, the automation recording is "latched" until playback stops. That is, the knob's or fader's position is continuously recorded, even if it is not moving, until playback ceases.

Note that multitrack mixers typically provide the ability not to only record automation in real time by moving real or virtual faders, but also the means for drawing the actual automation curves outside of real time using a pencil tool. Some software even offers elaborate curve-drawing options, such as functions to draw not only freehand but also with line segments, hyperbolas, Bezier curves, and so on. This allows the precise specification of automation parameters as well as the ability to fine-tune previously recorded information.

Besides simply storing recorded information about mix parameters for later playback, automation can serve as the basis for creative effects as well. A simple example is morphing between settings of an effects-processing plug-in. Consider an audio track that contains a sound file of a scream. The sound could be made to appear to move from near to far using a reverb plug-in using this technique. We could record the settings for a very small reverberant space at the beginning of the sound file, and then record the settings for a very large space at the end of the hall. By using either an automated parameter morph function or by manually connecting the appropriate automation settings curves, the physical space in which the scream occurs can be made to change, and hence we could perceive the virtual motion of the scream from near to far.

Macintosh-specific multitrack software. A popular Macintosh-exclusive multitrack workhorse is Digital Performer from Mark of the Unicorn (http://www.motu.com). Originally only a MIDI sequencing program called Performer introduced in 1980s, Digital Performer is now a full-fledged multitrack editor/sequencer that supports high-definition audio, surround mixing, a variety of plug-in formats, and sample-accurate synchronization. Additionally, Mark of the Unicorn produces an entire suite of hardware interfaces that integrate well with Digital Performer and other software as well.

Another excellent Macintosh multitrack editor/sequencer is Logic, from emagic (http://www.emagic.de). Previously available for both Macintosh and Windows operating systems, Logic has become a Macintosh-only program since Apple Computer acquired emagic in 2002. Logic too possesses a loyal fan base of long-time users, and it is particularly noted for its multilingual, customizable interface that allows rapid navigation. The program is also available in several price/functionality versions.

Other software for mixing on the Macintosh includes Deck and GarageBand. Deck, from BIAS (http://www.bias-inc.com), includes many features found on Digital Performer and Logic, and it is also available in a less-expensive limited version called Deck LE. A more consumer/"prosumer"-oriented program is GarageBand, from Apple, that includes loops, software instruments, and guitar amplifier modelers.

For those on a budget who do not require MIDI functionality, several low-cost alternatives for multitrack mixing are available exclusively for the Macintosh. These include the free program Cacophony by Richard Bannister (http://www.bannister.org), which supports up to 16 tracks and a wide array of sound file formats, and Jasmine by Kumilipo.com (http://www.kumulipo.com), which sports an attractive user interface and is available for well under US$100.

Windows-specific multitrack software. Three popular professional-level multitrack editor/sequencers are available exclusively for the

Windows platform: Sonar from Cakewalk (http://www.cakewalk.com), Audition from Adobe (http://www.adobe.com), and Samplitude from Magix (http://www.magix.net; http://www.samplitude.com). All programs command a large worldwide user base and include a wealth of professional-quality audio processing plug-ins.

Sonar from Cakewalk is a complete multitrack mixing environment that also includes facilities for loop-based composition and music notation. It supports both DirectX and VST plug-in architectures (we'll talk about plug-in architectures in Chapter 5), and MIDI plug-in effects. Furthermore, all mixing is performed using 32-bit words.

Adobe Audition is a 32-bit multitrack editor that supports sampling rates up to 10 MHz and includes a host of royalty-free samples and loops, audio effects and restoration plug-ins, and a surround-audio encoder. The customizable user interface is particularly compelling, and the built-in tempo identification and key-signature identification tools are especially helpful in many audio-for-video and other applications. Although Audition does not include MIDI sequencing and notation functionality, at least at present, it does integrate well with Adobe's other flagship media software, including Premiere and After Effects.

Magix's Samplitude does include MIDI functionality, although it is intended primarily for multitrack hard-disk recording, mixing, and mastering. It supports high-definition audio and incorporates a unique time-frequency audio display in its main window that assigns different colors to regions of audio based on their spectral content. For instance, when viewing a sound file of a vocal track, each middle-C note might appear green. This color-coding can save a great deal of time when editing sound files. Samplitude also includes a high-quality room simulator that uses convolution; we'll talk about how this and other similar room-simulators work when we get to plug-ins.

Another program, Maven3D (http://www.maven3d.com) from Emersys is a Windows-based multitrack editor specifically designed to mix and process surround recordings. While surround plug-ins are available for other multitrack editors, Maven3D includes proprietary technology for adding simulated but compelling three-dimensional space to mixes.

Linux-specific multitrack software.   Several high-quality multi-track editors are available for the Linux operating system. Although they tend to be somewhat less-integrated from a media-production standpoint (that is, most do not include sequencing or notation capabilities built-in), the fact that they are generally free and open-source makes these programs appealing to many users.

One of the most impressive software programs of any kind for Linux is Ardour from Linux Audio Systems (http://www.linuxaudiosystems.com), an open-source multitrack recorder/mixer available online at http://ardour.sourceforge.net. One of Ardour's most impressive features is that it performs all mixing internally using 64-bit floating-point numbers, and the total available number of audio tracks and busses is not limited by the program. Ardour also includes over 80 effects plug-ins and supports sample-accurate automation and unlimited undo/redo.

Other multitrack programs for Linux include the General Sound Manipulation Program (GSMP; http://drocklinux.dyndns.org/gsmp) by René Rebe, and ProTux (http://www.nongnu.org/protux) by an international team of developers headed by Luciano Giordana.

Cross-platform multitrack software.   At the center of an ever-increasing number of both professional and project studios is Pro Tools from Digidesign (http://www.digidesign.com). Founded in 1983, Digidesign has served as a central player in the development of the DAW, beginning with its launch of Sound Tools in 1989. Digidesign now offers a line of integrated software and hardware systems for turning Macintosh and Windows computers into professional audio workstations for recording, mixing, and post-production. Arguably the most full-featured multitrack editor/sequencer for any operating system, it can also be the most expensive. More often than not, professional recording and post-production studios are likely to contain at least one Pro Tools-equipped room, and Digidesign's newer line of project-studio-oriented systems are enabling budget-conscious users to work with Pro Tools software at home. One of the distinguishing features of most Pro Tools systems is that they offload most digital signal processing calculations to separate, dedicated hardware, enabling the host computer to concentrate on other tasks, such as file handling and user interaction.

Two cross-platform multitrack editor/sequencers are available from Steinberg (http://www.steinberg.de): Nuendo, intended for media production, and Cubase, intended for music recording and production. Both programs command a large international user base and feature customizable user interfaces.

Mixing: Where Are We Going?

As multitrack recording and mixing developed in the 1960s, a flurry of differing technical and aesthetic ideas regarding mixing audio tracks also arose. And one might say that, until recently, most computer software for mixing on the DAW has tended to emulate the working methodology and prototypes of the analog recording studio, and for good reason. We appropriate terms from the analog recording world all the time, because the concepts remain the same, and they work.

However, perhaps because the confluence among music, engineering, and computer science has now reached a certain level of maturity, new working paradigms involving the modern digital audio workstation have begun to emerge in both hardware and software incarnations.

The concept of the "channel strip"—that is, the set of control knobs, buttons, and fader that corresponds to a single channel of audio on a traditional or software recording console, is not necessarily the most creatively engaging interface for mixing music. Knobs, buttons, and faders may look prettier now than they did fifty years ago, but they are still knobs, buttons, and faders. Granted, a great number of people are extremely talented at working with the console in its current form, but is it the most efficient from a creative standpoint?

Asking this question is akin to the observation that the standard "QWERTY" computer keyboard that most people use is quite inefficient as well. It was, in fact, designed specifically to slow down typists so that keys in old typewriters would not stick together. The much-less-popular Dvorak keyboard, invented by August Dvorak and William Dealey in 1932, is more efficient in that it was designed so that the user's fingers naturally fall on the most frequently accessed keys. Despite its ergonomic advantages, it has never gained widespread acceptance. It is, however, loved by those who have learned and switched to it. Perhaps new paradigms for mixing and processing audio also will be popular, if among only a smaller number of users. Nevertheless, in the increasing extent to which the medium is the message and technology can either enable or stifle creativity, alternate mixing paradigms can play an increasingly central role in the power of the DAW.

Nowhere is this perhaps more apparent than with loop-based mixing software, such as Propellerhead Software's ReCycle (http://www.propellerheads.se), Sony Media Software's Acid (http://www.mediasoftware.sonypictures.com), and FL Studio (formerly known as Fruity Loops) from a company of the same name (http://www.flstudio.com), all of which are written for the Windows operating system. On the Macintosh, Soundtrack from Apple Computer (http://www.apple.com) includes many royalty-free sound samples and loops and is especially easy for novices to learn and use. In addition to these commercial programs, several shareware loop editors, such as ALLooper (http://www.allooper.com) for the PC, from Giammarco Volta and Giovanni Mazzotti, are also available.

Loop-based mixing software is optimized for specifically composing music that is based on loops or grooves. Tempo analysis, pitch shifting, and time-stretching algorithms are a crucial component of these kinds of programs so that loops that might not otherwise align in time or frequency can be made to "fit together." While loop-based mixers are quite popular and well suited for the stylistic norms of specific musical genres, speeding the creation of such music, they are often criticized for the very same idiomatic specificity that they place on the production process (as well as perhaps the ease with which bad music can be created!). But beauty, as they say, is in the eye of the beholder; if the tool is not right for you, then use something else.

Prototype alternative hardware controllers for mixing are an active area of research. Consider, for example, the AudioPad, developed by James Patten and Ben Recht at MIT, which assigns tracks of audio to physical "pucks" that a user moves about on a small table to control relative amplitude, spatial position, effects processing, and other parameters. Other mixing controllers allow users to move tracks of audio about in three-dimensional space, or to mix audio using wearable devices like gloves.

Several commercially available software programs also creatively and effectively re-examine the standard track–channel mixing prototypes found on most multitrack editor/sequencers available today. These include Ableton Live (http://www.ableton.com), which is particularly useful for real-time mixing applications; Melodyne from Celemony Software (http://www.celemony.com), which displays spectrograms of audio tracks and pitch/tempo information rather than time-domain waveforms; and Reason from Propellerhead Software (http://www.propellerheads.se), which incorporates a virtual sequencer, drum machine, mixer, synthesizer, and sampler into a unique integrated production environment.

What will the virtual mixer of the future resemble? No one knows precisely, but I submit that it, like most software of the future, will engage the user more directly by diminishing the user's awareness that software, or even a computer, is being used. Thanks to the potential for creative human–machine interface technology, the bridge between intention and sounding result will be made as short as possible.

Will we be using a computer mouse or hardware faders on a mixing console in the future to mix music? Sure—many, if not most, people will. But I hope that we will tangibly sculpt sound in a more direct manner.

Traditional Mixing Techniques

Acquiring the fundamental techniques and artistic skills required to craft an elegant mix is a formidable task, as is even attempting to talk about mixing in general terms. And the bad news (or the good news, depending on your perspective) is that the very prospect of working with music technology—or rather technology in general—ultimately relegates one to a lifetime of student status. Staying current in any field that relies heavily on technology requires frequent re-tooling as newer and often better software and hardware becomes available. Thankfully, the skill set one acquires is usually transferable to current technology, particularly in the case of using the DAW for mixing, as the fundamentals of mixing technique remain more or less the same. As you develop your own voice, you should be able to express that voice regardless of the tools you use to mix, unless your craft and voice are tied (either intentionally or unintentionally) to a particular piece of technology. (For example, some mix engineers insist on using a particular piece of their favorite gear or a certain plug-in for virtually every mix.)

As I mentioned, it is incredibly daunting to discuss mixing in general terms, because what works for mixing one particular style of music may not apply at all to another style. Nevertheless, psychoacoustically explainable phenomena common to all kinds of music, such as "muddiness," "boominess," phase problems, and masking issues allow some generalities to be drawn.

At its most fundamental level, mixing involves five steps:

1. Selecting sound files to mix
2. Time-aligning them in the desired way
3. Adjusting their relative amplitudes

4. Adjusting their spatial locations
5. Effects processing

And while we can aesthetically reduce mixing to these five elements, the technique involved in effectively executing each one can take a great deal of practice and time. (The techniques we discuss in this section will develop steps 2 through 5. The first step involves either recording or synthesizing sound files, and we cover those topics in Chapters 8 and 9.)

The traditional mixing techniques discussed here are, in general, not specific to the DAW. However, owing to its inherent robustness and flexibility, the DAW can offer an enormous amount of control over the way some of these techniques can be implemented. And some tools that are useful for implementing these techniques, such as spectral-domain noise removal and a great many effects processors and plug-ins, are simply not available outside of the DAW.

Before beginning, consider two different approaches to mixing: *less is more*; and *more is more*. Which overriding approach is best? Here are two simple scenarios to explain what I mean.

- Will the application of a large number of digital audio effects to a vocal track make it more intelligible? Will automatically retuning the slightly out-of-tune notes increase or decrease the emotional expressivity of the performance? If preserving intelligibility and maintaining the original emotion of the voice is the goal, then perhaps "less is more."
- In a very thick mix, perhaps the goal of the composer or mix engineer is some kind of overload of sonic complexity. It might be desired to create a climax of emotional density by pulling out all the stops, so to speak. In this case, we might say "more is more."

Even though the DAW offers a powerful environment for mixing and allows straightforward implementation of both of these mixing paradigms, I have found in my experience that beginners fall almost exclusively into the "more is more" camp, even if the mix doesn't warrant it. Consider carefully the roles that clarity, intelligibility, and economy might play in expressing your musical ideas, and fight the temptation to believe "more is more" in order to cover up a not-so-great recording or synthesized track, which leads us to the first traditional mixing technique.

Starting with Quality Ingredients

Whether a mix involves recorded tracks, synthesized tracks, or a combination of the two, using only the best quality source material increases the likelihood that the mix will work musically and can save many headaches down the road. Smoothing over a rough track by "fixing it in the mix" frequently does not lead to a quality mix! I have occasionally tried to incorporate a recorded or synthesized sound file into a mix that I was not completely satisfied with on its own terms, and I would invariably end up wasting time trying to make it fit within the mix, only to throw it out later. Spending the extra time to create sound files and recordings that "work" on their own terms—sounds that are free from excessive undesired noise, distortion and the like, and that you are happy with by themselves—will almost always save time in the long run.

This concept harkens back to the previously discussed paradigm of editing sounds first in a two-track mixer if possible before importing them into a mix. There are at least two advantages to this. First, we can apply necessary equalization, noise reduction, and other processing in a two-track editor, thereby reducing processor load when mixing them in a multitrack editor-sequencer. Applying real-time effects to tracks in a multitrack editor-sequencer can dramatically tax the computer's central processing unit, potentially limiting the number of available tracks and the number of effects that can be applied. Second, editing sounds first within a two-track editor will help ensure that we are starting with "quality ingredients" in the mix. A mix engineer should be primarily concerned with creating ensemble, clarity, space, and dynamic shaping—not with musically repairing the deficiencies in individual tracks.

The Rough Mix

To get started in creating a mix, many engineers suggest creating a "rough mix" as quickly as possible. Assembling recordings and sound files in an approximate temporal alignment can help provide the "big picture," and doing so quickly can ensure that the global representation of the mix does not become mired in details. Indeed, beauty is in the details, but the rough mix can provide the rough canvas on which to assemble these details.

Comparative Mixing

Many mix engineers find it useful to occasionally refer to other well-known mixes in a similar musical idiom during a mix session. That is, they periodically interrupt their session to listen to a "reference mix." (This can be effective whether or not you yourself made the reference mix; a well-mixed, commercially available recording whose features you like should do just fine.)

Comparative or reference-based mixing can yield insights regarding whether your current mix is congruent with certain accepted stylistic norms, particularly when mixing commercial ("popular") music. For example, listening to a reference mix in the middle of a session might let you know that your mix is considerably quieter overall than another mix in a similar idiom. This might suggest compressing and/or applying gain to your mix. Listening to a reference mix might also provide insights into equalization issues that should be addressed in your mix.

Listening to reference mixes can lead to greater consistency across mixes, and it can particularly helpful when mastering an entire album. On the other hand, it may be less applicable to noncommercial experimental music in which no similar reference mix exists.

Mixes can also be compared using feature extraction, which some engineers and composers (particularly the more technically inclined ones) find quite useful. For example, a friend of mine always compares the RMS amplitude value of all the samples in a new mix to the RMS amplitude of other mixes he has made, and he attempts to achieve a consistent value as much as possible. You could also compare mixes using spectral feature extraction if you are so inclined. For example, comparing the spectral centroids[5] of a new mix to a reference mix would indicate the relative "brightness" or "darkness" of the new mix. This kind of feature extraction can be performed via custom software or with some commercially available analysis plug-ins.

Multiple-Monitor Mixing

It is often said that it really does not matter what loudspeaker monitors you use, but rather how well you know them. And there is some

---

[5] The spectral centroid of a collection of audio samples can be calculated according to a simple mathematical formula. It tells us the weighted mean of the sound's spectrum. Low centroids indicate that more spectral energy is concentrated in the bass frequencies, while high centroids suggest the energy is concentrated in treble frequencies.

truth to that statement, particularly among high-end monitors. They of course vary in many ways: frequency response, transient response, spatial imaging, and more, but developing and maintaining an understanding of how they sound musically is of the utmost importance to creating a successful mix.

Studio monitors are unlike consumer stereo speakers because they are especially designed to not imbue the sound they reproduce with any artificial coloration. That is, they are designed to respond equally to all frequencies within as wide a range as possible. Consumer speakers are usually designed to respond better to low bass and high treble frequencies than to midrange frequencies, which tends to make popular music sound better according to most people's tastes.

An exception to this line of thinking, however, can be found in the immensely popular Yamaha NS-10 studio monitors. These were originally introduced in the early 1980s as consumer bookshelf stereo speakers, and their frequency response is not particularly stellar—actually far from it. It is often said, however, that any mix that sounds "good" on a pair of NS10s will sound "good" on just about anything! You might say that this is the "mixing to the "least common denominator" approach.

Whatever primary monitors you use, mixing with multiple monitors can help ensure that a mix is not dependent on the characteristics of any one loudspeaker. And it is crucial to listen to a mix on the intended playback system as much as possible. Is a mix designed to be heard primarily over the radio in a car? Then listen to it in a car, and take it back to the studio to refine it. Is a mix designed for concert playback in a large hall? Listen to it in the hall, or measure the impulse response of the hall and convolve it with your mix to hear a virtual mock-up.

As a side note, bass frequencies present perhaps the most difficult-to-control variable during playback of a mix. Clearly, a basic car stereo will treat bass much differently than a home theater with a subwoofer. Listening to a mix on the widest variety of playback systems and monitoring on different speakers can help prevent surprises later and lead to more effective treatment of bass.

Low-Level Monitoring

It is recommended that peak mixing levels in a control room generally do not exceed 85 dBSPL. (You can check your levels with an inexpensive SPL meter.) Quieter sounds are easier on the ear than louder ones (particularly for long periods of time), and mixing at softer levels helps reduce ear fatigue. It can also save your hearing! Furthermore, the mechanisms of the ear can lead to distortion when presented with loud sounds, which can yield an inaccurate sonic picture of the mix.

In addition, psychoacoustic studies show that, all things being equal, most think that "loud" sounds "better." It follows quite simply that if your mix sounds good at a softer level, it should sound even much better when played back louder!

Making Multiple Mixes

Making several versions of a mix can be helpful. With commercial music, providing multiple mixes enables a mastering engineer to choose the most appropriate mix within a larger context for inclusion on an album or compilation. Some suggest producing several mixes of a song with varying degrees of effects processing, bass levels, and vocal levels.

With concert tape music, the resulting mix might well be played in various venues, from small, dry black-box theaters to large, airy concert halls. Having wet and dry mixes of a composition available for a performance greatly increases flexibility when dealing with different performance venues.

When making multiple mixes, it may be helpful to allow time to pass between versions. Listening to the sonic material with a fresh set of ears after a few days away from it can be very inspiring.

Controlling Noise

Three primary methods for removing noise are available: physically deleting or gating noisy sections of a track, expanding a track's amplitude levels in the time-domain, and processing a track with spectral noise-reduction software. By trial and error, you can develop an intuition for what is most appropriate in different contexts, but whatever the method, removing unwanted noise is crucial to the resulting clarity and intelligibility of a mix. Although tracks played back by themselves may not sound very noisy, many such tracks can create quite a bit of resulting noise when mixed together.

I do not mean to say that all noise should be removed, simply all unwanted noise. Noise is entire aesthetic territory unto itself, and many successful mixes employ what we normally think of as "noise" in creative and useful ways. If a track contains a noninformation-bearing or undesired signal ("noise"), it may be worth eliminating for the sake of clarity. If "noise" is central to a musical or other aesthetic idea, then it is information-bearing, and hence axiomatically not noise.

If you do employ noise-reduction software or hardware, employing a constant ear toward any processing artifacts is crucial. I have heard mixes in which the tracks may be free of extraneous noise, but the FFT "gurgle" by-products that sometimes result from overapplying a noise-reduction plug-in are quite distracting.[6]

Tuning Problems

The modern DAW offers a robust collection of tools for correcting tuning problems within a mix. And to be sure, these tools, when used appropriately, can be very effective and tonally transparent. However, many mix engineers suggest avoiding overuse of tuning correction software, which can lead to spectral coloration and may remove the emotional, affective, and intentional pitch slides and blue notes from a track when not used with the utmost care. It is far better to record a track as many times as necessary to yield a fully in-tune performance than to "fix it in the mix."

Critical Bands

Our auditory processing systems tend to dissect acoustical stimuli into slightly overlapping frequency channels. Acoustical phenomena that occur within each frequency channel are processed using more or less the same neural mechanisms, which may vary somewhat from channel to channel. These frequency channels are called *critical bands*, and it is commonly accepted that there are about 24 critical bands over the range of human hearing. Music cognitionists have proposed various mathematical formulae to model the bandwidth of each critical band as a function of frequency, based on extensive testing of human subjects.

---

[6] Any processing algorithm that uses the Fast Fourier Transform (FFT) to analyze and then resynthesize a signal, as many noise-reduction algorithms do, is subject to audible artifacts.

According to Rossing (1990), very tiny sections of the inner ear's basilar membrane are tuned to respond to each critical band, and each of these sections is tied to around 1,300 neurons that directly feed auditory information to the brain. Some research has found the frequency width of each critical band is approximately 30 times greater than the *just-noticeable difference* (JND) of each frequency. (The JND of each frequency is defined to be the empirically determined frequency difference that the "average" listener can discern half the time relative to that frequency.) Two frequencies occurring within a critical bandwidth are more likely to be perceived as a single "rough" or "beating" frequency than as two distinct frequencies. Once the frequencies are at least one critical band apart, then we begin to perceive them as two separate tones. The critical band frequency cutoffs are codified by the International Standards Organization (ISO) as the *Bark scale*, and they are shown in Table 4.2 defined up to 15.5 kHz.

TABLE 4.2

The ISO Bark Frequency Scale

Bark Number	Low-Frequency Cutoff (Hz)	High-Frequency Cutoff (Hz)	Bandwidth (Hz)
1	0	100	100
2	100	200	100
3	200	300	100
4	300	400	100
5	400	510	110
6	510	630	120
7	630	770	140
8	770	920	150
9	920	1,080	160
10	1,080	1,270	190
11	1,270	1,480	210
12	1,480	1,720	240
13	1,720	2,000	280
14	2,000	2,320	320
15	2,320	2,700	380
16	2,700	3,150	450
17	3,150	3,700	550
18	3,700	4,400	700
19	4,400	5,300	900
20	5,300	6,400	1100
21	6,400	7,700	1300
22	7,700	9,500	1800
23	9,500	12,000	2500
24	12,000	15,500	3500

An understanding of critical bands is crucial to successful mixes, because musical information occurring within a particular critical band can get "crowded." One of the psychoacoustic phenomena that occurs within critical bands is *masking*, which occurs when a tone or frequency band hides behind ("masks") another. Frequency masking is most likely to occur within one critical band, so the more frequencies that are present within a critical band, the more likely one or more of them is to be masked.

Owing to masking and other effects, there is a nonlinear nature to how we process frequency information. As such, we say that *pitch* is the psychoacoustic correlate of frequency.

To summarize: Don't crowd the critical bands. This axiom can inform the way we use equalization and help us pick the kinds of sounds that best fit together psychoacoustically.

Fletcher-Munson Curves

A friend of mine calls these the "curves to live by." In 1933, Harvey Fletcher and W. A. Munson wrote a paper in which they measured the sensitivity of a number of human subjects to different sine-tone frequencies. The results they obtained (Figure 4.20) have since been slightly modified (notably by Robinson and Dadson in 1956) and are codified as a specification by the International Standards Organization (ISO).

There is a nonlinear nature to how we process amplitude information in audio signals. As such, we say that *loudness* is the psychoacoustic correlate of amplitude. Loudness of sounds is measured in units of *phons*, while the amplitude of sounds is measured in units of atmospheric pressure (bars or pascals). Once acoustic pressure waves are transduced into electrical signals with a microphone, we can then measure their amplitudes in units of volts; once they have been digitized, we can measure their amplitudes in units of quantization levels.

Why are these the "curves to live by" when mixing? A working knowledge of these curves coupled with a good equalizer plug-in can help a mix to counteract the nonlinear, frequency-selective nature of our hearing. For example, because our ears are more sensitive to 4 kHz than 80 Hz, a little EQ can go a long way at 4 kHz, while much more may be needed at 80 Hz to achieve a comparable loudness change. These curves are important to consider when using any kind of effects processing that can alter the tonal balance of a sound.

Figure 4.20
*Equal-loudness
contour curves,
after Fletcher and
Munson (1933).*

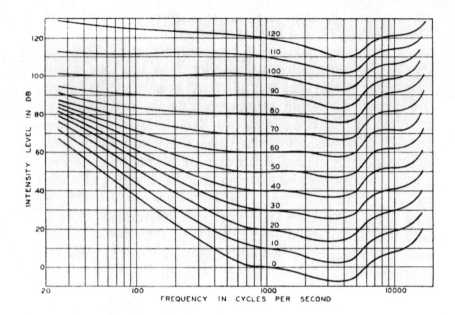

Effects Processing

The most often-heard advice given with respect to effects processing in a mix is simply "don't overdo it." The fact that a DAW might have the processing power to apply reverberation, dynamics, equalization, pitch-shifting, and time-scale expansion to each individual track in a mix might make it tempting to push the system to its processing limits, but this might not make musical sense!

Before applying any effect to a sound, it is crucial to consider these questions:

- What effect will help this sound speak musically?
- What effect will give a sense of life, of organicism, to this sound?
- What effect will this effect have on the intelligibility of the sound?
- How will this effect impact the way this sound interacts with other sounds in the mix?
- Is this effect congruent within the musical logic of the mix?
- Is this effect necessary?
- Why should I use this effect instead of another?

Again, consider creating multiple versions of a mix with varying degrees of effects processing. You can always return to pick your favorite at a later date.

Vocal Intelligibility

With most commercial music featuring a vocalist, it is generally considered desirable to hear and understand the vocalist's words and utterances Of course, if the vocalist is not singing words but rather vocalizing syllables, for example, intelligibility is clearly not an issue. But to the extent that intelligibility conveys a desired semantic meaning, it is of the utmost importance in the mix. Again, listening to your mix on many different monitors can help ensure vocal intelligibility among a wide range of loudspeakers upon playback.

Even in music featuring vocalizations ("oooh–ahhh") or extended vocal techniques ("ssszzzdddzdggggvwmth–t–t–t–t"), the voice should be treated with the utmost care in the mix. In his book *Audible Design*, the composer Trevor Wishart points out the importance of the vocal "utterance." For various physiological and evolutionary reasons, we are drawn to the recognizability of the voice across all cultures. This immediacy of experience, and the resulting way in which we listen to and perceive the human voice in a recording, dictates that we must give the voice special priority or consideration, or that we must at least consider treating it in a special way, in a mix. Just how one defines "priority" or grants "consideration" is up to the individual mix engineer to decide.

Dealing with Drums

A problem sometimes encountered when recording a live drummer is inconsistency of attacks, particularly in snare drums, as it is enormously difficult to strike the snare with identical force each time. For that matter, percussion instruments in general can present a mixing challenge owing to their enormous dynamic range. (A loud cymbal crash or canon explosion, as in Tchaikovsky's *1812 Overture*, can certainly be difficult to mix!)

To alleviate this inconsistency of attacks, especially in groove-oriented drum tracks, mix engineers frequently replace noticeably quieter attacks with copies of other previously played attacks by simply cutting and pasting. After replacing the offending attacks with samples of attacks at a more consistent level, mix engineers also often compress the dynamic range of the percussion, sometimes quite radically. Compressing the percussion before replacing the offending attacks may be a disaster, however, because the amount of compression required to make everything sound consistent in amplitude might lead to noticeable overcompression and so-called "pumping and breathing" audible by-products. Careful and judicious application of compression, however, can lead to somewhat tighter tracks that are less muddy.

Phase Problems

Does phase impact the way we perceive sound? It's a sticky question, and the answer is somewhat context-dependent. Consider two examples. The first, illustrated in Figure 4.21, is a square wave, which can be decomposed into a Fourier Series of an infinite sum of odd harmonics of a single sine wave (with the nth harmonic weighted by a factor of $1/n$). Figure 4.21a shows a time-domain representation of a square wave as we are accustomed to seeing it: It does indeed look very square and boxy. This particular square wave was computed by adding the first 1,000 odd harmonics of 220 Hz. (I removed the axes for clarity.) But what happens if I alter the phase, or starting times, or each of the component harmonics? By simply altering the phases of

Figure 4.21

(a) Time-domain representation of a square wave; (b) another time-domain representation of a square wave; (c) yet another. Only the phases of the partials have been changed.

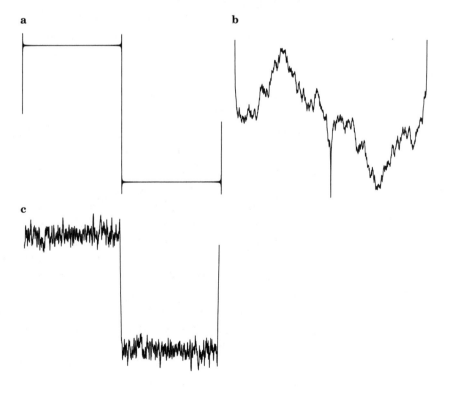

the harmonics, I can arrive at Figure 4.21b, which looks almost nothing like Figure 4.21a. In Figure 4.21c, I randomized the phases again, and we get another different picture.

Here's the funny part: all three square waves sound the same! In general, we can't tell them apart, so we could easily conclude that phase plays no role in our perception of sound.

But we would be incorrect. While the three square waves played in isolation sound identical, differences can be heard when an amplitude envelope is applied. For example, if we apply a short percussive envelope to the two square waves so that they only become 100 millisecond bursts, the phase differences can definitely color the way we perceive the transient nature of the sound (because the initial arrival of the harmonics are reaching our ears at different times) as well as the way we perceive their spatial location. (We talk more about this later.)

Some phase problems can be simply fixed by simple operation: most mixing consoles, and hence multitrack editor/sequencers for DAWs, include a "phase-invert" button for each track. By muting and soloing various track combinations, you can use your ears to find the phase problem, and it can often be alleviated by simply inverting the phase of the offending track.

Phase problems can mean the difference between a tight, punchy mix and a duller, muddy one. And a thorough understanding of phase can be particularly critical when examining our next topic: panning.

Panning Issues

The first question many seasoned mix engineers ask themselves while working on a mix is simply, "Does it work in mono?" That is, if you downmix all of the output channels to one channel (i.e., pan them all to one loudspeaker), does the mix behave as you expect? If it sounds substantially different when downmixed monaurally in terms of frequency content, brightness, and transient qualities, then phase problems have most likely coupled with panning problems to muddy the mix. First, try to correct the phase problems as described above, and then adjust the panning of tracks until a musical solution is found.

Mix engineers differ greatly in their panning techniques. Some prefer to pan everything hard left, hard right, or center, while others create a more continuous sound stage. Whatever your aesthetic direction, it is important to not use panning as an aural gimmick. Spatial location is of course one of the crucial elements in our perception of sound and can be an extremely important compositional and mixing parameter. But panning for its own sake and not a greater artistic purpose can quickly sound like a novelty.

The overriding principle to consider when using panning is whether you are treating the loudspeaker as a point-source of sound or as a virtual window into acoustic space. This becomes particularly important in surround mixing, in which we have more loudspeakers to worry about. But the idea is this: Are you trying to acoustically convince the listener that the speaker is the literal object which it portrays, or is it trying to provide a glimpse of a created acoustical reality?

Effective Equalization

The tonal balance of a mix can quite simply create the difference between a well-balanced, pleasing overall sound and an ear-fatiguing, mushy quality. It takes practice, and an understanding of the Fletcher-Munson equal-loudness curves, to learn how to equalize effectively for a given musical style, but one of the style-independent techniques often cited is subtractive equalization.

Equalizing the tonal balance by subtracting frequencies rather than boosting them is useful for two reasons. First, subtractive equalization keeps the noise floor lower in general. Even in floating-point digital systems, amplification raises the noise floor and eats into the available headroom. Second, subtractive equalization is more likely to correct destructive interference between frequencies. If two adjacent frequency bands are interfering, causing a diminished intelligibility and muddiness in the region, then amplification of one of the bands would not effectively address their interference: it would just make one of the frequency bands sound louder. If the bands are interfering, then equalization by attenuating one of the bands can effectively minimize their interference.

It's also a good idea to practice identifying frequencies. Getting to know what 40 Hz, 800 Hz, and 8 kHz (for example) sound and feel like to your ears can greatly increase your fluency with equalization techniques.

As a side note, equalization should not be used to compensate for poor microphone placement. We have not yet spoken about basic recording techniques, but keep in mind that proper microphone placement when recording is far better than fixing any problems later with equalization.

Space

In addressing issues of spatial location and trajectory within a mix, we could simply list various recipes, for example "mix everything hard left, hard right, or center" or "always keep vocal sounds panned dead center." But the issue is much more complicated and very much context-dependent. And now that surround mixing is becoming more popular (and affordable), an entirely new dimension is available for creating and controlling space.

Effective use of musical space requires implicit control of three dimensions: the perceived distances of sound sources from the listener in the horizontal (left-right), transverse (near-far), and vertical (top-bottom) dimensions. These can all be controlled more or less independently, to some extent, regardless of the delivery format of the mix (stereo, 5.1, 7.1), although the vertical dimension is the most difficult, unless you can afford floor and ceiling monitors! There are some nifty tricks available to help us create spatial illusions, though.

Different psychoacoustic processes and physical phenomena are involved in helping us achieve this goal, and I think this area is so important, we'll spend the entire next chapter covering it. Stay tuned!

Learning by Doing

From an aesthetic perspective, mixing is clearly an art, and there are no clear-cut recipes for success. Although recipes and free advice are readily available, much of the best music ever composed and mixed pushes the musical envelope of its time, frequently defying convention in its quest for new sounds, expanding our notion of what music is and can do.[7] Remember that in the distant past of Western musical history, harmony itself was considered by many to be either too weird or simply sacrilegious; only voices singing in unison were allowable. And when harmony did first appear, it was what we might now call "simple": parallel octaves and parallel fifths primarily. The trajectory from that mode of thinking to the present use of electronics and com-

---

[7] With all that said, I would like to point out an excellent resource for further exercise and study, William Moylan's *The Art of Recording: Understanding and Crafting the Mix*. The discussion and practice exercises therein are a great resource for learning more about how to mix.

puters in the creation of virtually all music heard today is one that clearly points forward, embracing an ever broadening conception of music created by sound itself.

I am not trying to argue solely the virtues of experimentalism over tradition in art: wonderful, relevant, beautiful, powerful music has been written by both experimentalists and traditionalists in all periods of music history. As the Italian composer Ferrucio Busoni (1866–1924) noted, "the old and the new have always existed, and will continue to exist." Rather, I am trying to say that great music—and great mixes—can be made using either traditional mixing techniques and practices like those briefly outlined in the preceding sections or potentially by completely ignoring them and making up your own.

With that said, it is obvious that simply using traditional techniques or inventing your own techniques does not guarantee a great mix. It takes a great deal of time and practice, and even the most well-respected mix engineers still hone their craft and technique daily. As a classically trained composer of computer music, I personally tend to resonate with mixes that defy convention in creative ways, not for the sake of defying convention itself, but because they explore new worlds of sound otherwise unavailable. But I recognize that this perspective is not for everyone, and that clearly, new sound worlds and unexpected musical landscapes can be created using traditional techniques as well.

You don't have to be a strict traditionalist or an experimentalist when it comes to mixing; perhaps some of the basic techniques work for you and others do not. The point is that the only way to know is to learn by doing. Mix as much music as you can. Mix as many different styles of music as you can, especially when starting to learn the craft of mixing. And read the other books that offer "recipes" for mixing for perspective. As with most fields, it is easier to successfully break convention if you first know what convention is and even master it. Bach and Beethoven wrote some really great music by following some conventions, ignoring others, and creating new ones altogether.

The foregoing discussion is predicated on an understanding of mixing as a form of music composition. One of the traditional mixing techniques previously discussed was to make many different versions of a mix. Clearly, one could easily create mixes of an identical song or composition that, although they used identical sound files and musical source material, are so vastly different that they mask their common source—that they are perceived as entirely different songs or compositions. This points to the crucial role that mixing plays in shaping a song or composition. In fact, the extent to which a mix shapes our perception and aesthetic understanding of musical intention is the very same extent to which mixing is itself music composition.

For Further Study

Fletcher H, Munson WA. "Loudness, Its Definition, Calculation and Measurement." *Journal of the Acoustical Society of America* 1933;5:82–108.

Knave B. "Ten Tips for Nailing a Mix." *Electronic Musician* July 1, 2000.

Knave B. "Mysteries of Mixing." *Electronic Musician* April 1, 2001.

Moorefield V. "From the Illusion of Reality to the Reality of Illusion: The Changing Role of the Producer in the Pop Recording Studio." Ph.D. Thesis, Princeton University; 2001.

Patten J, Recht B, Ishii H. "Audiopad: A Tag-based Interface for Musical Performance." *Proceedings of Conference on New Interface for Musical Expression* (NIME '02). Dublin, Ireland, May 24–26, 2002.

Phillips D. *The Book of Linux Music and Sound.* San Francisco: No Starch Press; 2000.

Robinson DW, Dadson RS. "A Re-Determination of the Equal-Loudness Contours for Pure Tones." *British Journal of Applied Physics* 1956;7:166–181.

Shea M, Everest FA. *How to Build a Small Budget Recording Studio from Scratch.* New York: McGraw-Hill; 2002.

Steiglitz K. *A Digital Signal Processing Primer.* New York: Addison-Wesley; 1996.

Exercises and Classroom Discussion

1. To end a mix, we could either bring the sound to an abrupt halt (with a crescendo, for example), or we could gently fade out into silence. List some other techniques for ending a mix, and try them all on a mix on which you are currently working.

2. Select two different currently available multitrack editor-sequencers, and compare and contrast their features.

3. What is the best mix you have ever heard? Play it for the class and explain why you think it works.

4. What is the worst mix you have ever heard? Play it for the class and explain why you think it does not work.

5. A common way to transition or "morph" from one sound to another is with a simple cross fade. How else could you morph between sounds?

6. What does the following filter do?
$$y_t = \tfrac{1}{3}(x_t - x_{t-1} - x_{t-2})$$

7. What effect would the following filters have when applied to a sound file whose sample rate is 96 kHz?
   a. $y_t = x_t$
   b. $y_t = x_t$
   c. $y_t = x_t + x_{t-96000}$
   d. $y_t = x_t + 0.5 \times y_{t-100}$
   e. $y_t = x_t + x_{t-96000} + 0.5 \times y_{t-96000}$
   f. $y_t = x_t + 0.9999 \times y_{t-100}$

8. What does a 50-day moving average of the Dow Jones Industrial Average really tell you? Would sudden changes in the market be more or less apparent by viewing a 20-day moving average or a 50-day moving average?

9. Experiment with a mix on which you're currently working in the following ways.
   a. Pan everything either hard left, center, or hard right. Compare the image and localization of parts relative to the existing mix.
   b. Pick a track and distort it in various ways.
   c. Listen to your mix in a car and on a boombox. Does it sound fundamentally different than it did in the studio?
   d. Listen to your mix in mono. Do not simply mute the left or right master track; pan them both to center, and see if the mix still works.
   e. Run the master output through a 20:1 limiter. Run the original mix through a 2:1 compressor. Comment on the differences in the 20:1 limited version and the 2:1 compressed version.
   f. If you are working with pitched material, is everything in tune relative with each other? Why or why not?
   g. Does your mix sound just as good at much quieter levels? Use a sound level meter to ensure your are not listening to levels higher than 85 dB.

10. Construct a mix entirely out of one sound file.

11. Play with a delay/echo plug-in in the following ways:
   a. What happens when you set the delay time to a really small time, like 1 millisecond? What about 2 milliseconds? Can you find a relationship between the delay time and frequency ratios that you hear?

b. What happens when you set the feedback gain on an echo plug-in higher than 100 percent, and why?

c. How might working with a surround audio mix introduce new parameters to control with a delay/echo plug-in?

d. In this chapter, I listed a simple filter equation that illustrates mathematically how a delay plug-in works. Create a corresponding filter equation for an echo plug-in, keeping in mind that it is a feedback filter.

12. Listen to the composition *Piano Phase* by Steve Reich, a classically trained composer who is often called the "father of techno."

a. Experiment with looping with your DAW to create a similar "phase" piece out of different audio material.

b. The "phase" in *Piano Phase* refers to two identical loops that are played slightly out of tempo phase with respect to each other. What other elements of music aside from tempo might be subject to this kind of "phasing" technique? Write a short musical study to illustrate your point.

13. Would you agree or disagree with the statement that the RMS amplitude of popular music mixes has increased since the introduction of the compact disc in the early 1980s? How could you prove or disprove this?

Plug-Ins

Whatever the audio software of tomorrow resembles or the techniques it embraces, a concept ushered in by the multitrack mixing programs of the 1990s that is sure to stay with us for awhile is that of the plug-in, to which our attention now turns. A plug-in is a piece of software that adds to the functionality of a larger, existing program. The plug-in itself defines an audio process, like displaying a volume level meter of the audio on which it is applied, or perhaps applying a reverberation algorithm to an audio track. A plug-in cannot run by itself in general; it requires a host to run it (and necessarily to support the format in which it was written, of which there are many as we will see shortly).

Arguably, software plug-ins are second only to multitrack editor-sequencers in terms of the musical (and economic, for that matter) value they bring to the DAW. The plug-in version of a hardware processing box may cost considerably less than its hardware counterpart. For example, the Waves L2 Ultramaximizer plug-in costs around US$500, while the hardware version costs over US$2,000. A hardware sampling reverberation unit from Sony like the DRE-S777 can cost over US$7,000, while the similarly featured Altiverb plug-in from Audio Ease costs US$495.

Additionally, because plug-ins exist as software rather than hardware, they are much less likely to need the repair that hardware boxes sometimes do! Like their hardware cousins, though, they can be stolen, which has led most manufacturers of plug-ins to implement various copy-protection schemes.

We concentrate first on some of the technical aspects of plug-ins, followed by a discussion of the aesthetics of musical effects in general. We conclude by addressing the specific architectures of various plug-in formats.

Families of Plug-Ins

Plug-ins can be categorized in many different ways. We can talk about them in terms of their musical applications (which plug-in is applicable to which musical scenario?), their method of operation (time-domain processing, frequency-domain processing, etc.), their price (some are free while others are far from it). Plug-ins can also be characterized by the technical goal they help the user accomplish. The three broad categories of plug-ins are *sound analyzers, effects processors*, and *virtual instruments*. Sound analyzers attempt to inform the mix engineer of some important sonic feature, such as the spectral content or spatial imaging of a mix. They do not alter the sound to which they are applied. Effects processors, on the other hand, do alter the sound to which they are applied. They often employ some sort of feature analysis on sound just as sound analyzers do, but the end result is a processed version of the sound to which they are applied. Virtual instrument plug-ins, like their standalone application counterparts, include software synthesizers and samplers that either model the sound of an existing musical instrument or allow the user to create new sounds altogether.

We survey each of these types of plug-ins based on the specific technical goal they help the mix engineer accomplish while examining some of the more popular ones in greater detail. We then discuss some of the musical ways they can be used. Note that some of the kinds of plug-ins outlined in this section are occasionally included as built-in features on some DAW software. It seems that today's plug-in is tomorrow's standard feature on many software programs.

Sound Analyzers

Sound analyzers can be broken down into six primary categories: time-domain analyzers, frequency-domain analyzers, spatial imaging analyzers, tempo detectors, pitch detectors, and experimental analyzers. Learning to effectively use information provided by sound-analysis plug-ins can help in two ways:

1. Sound analyzers can help you achieve consistency from one mix to another. For example, when mastering an entire compilation, it may be desirable to ensure that each song or composition exhibits roughly the same peak amplitude.
2. Sound analyzers can provide information about sonic features that are either difficult or time-consuming to measure manually or by ear. For example, an accurate tempo detector, useful in tempo-locking different tracks and selecting drum loops to import into a mix, may save time over measuring tempo manually with a stopwatch or trying to synchronize a groove to a metronome. They can also be more accurate than the human ear.

Time-Domain Analysis

Audio signals can be analyzed and dissected in terms of their temporal feature content (for example, what the amplitude of the signal is at a particular point in time), their frequency feature content (for example, what overtones are present in a sound and in what amount), and in terms of their combined time-frequency content. Simply enough, analyzers that work on signals purely in the time domain, by looking at their sample values as a function of time, are called time-domain analyzers.

Time-domain analyzers are most often used to display the amplitude of an audio signal, which is very helpful when setting levels in a mix, but they can do this in several modes, usually called *VU metering* (Figure 5.1), *RMS metering*, and *peak metering*. VU (volume unit) meters were designed for analog recording and mixing systems, and their characteristic "retro" analog appearance is often incorporated into DAW plug-ins. They display a kind of "average" level of an audio signal. More precisely, VU meters attempt to display a level corresponding roughly to the perceptual correlate of amplitude: loudness. To this end, VU meters incorporate a lag in their display: transient information, like the quick burst of a piano attack or the pluck of a guitar string, is generally too fast to make the VU meter rise instantly. In fact, transients less than about one-third of a second register a very accurate level, because the meter is designed to respond to average signal levels. VU meters are somewhat less useful in digital audio systems than in their analog counterparts, because the built-in lag can prevent them from indicating overloaded transient signals, like an occasional "hot" burst in a signal that clips. Remember from Chapter 1 that, owing to their respective distortion characteristics, analog systems tend to mildly round the edges of overly "hot" signals, whereas digital systems clip them to the outer rails. This is partly why an occasional overloaded burst on analog tape is often allowable, but never so in a digital signal. Nevertheless, analog VU meters are often emulated in software plug-ins, and many mix engineers rely heavily on them to double-check their ears.

RMS meters are similar to VU meters in that they display a kind of "average" signal level and give a better idea of the perceptual loudness of an audio signal, but they are calibrated a bit differently. Taking the RMS value of a continuous, analog signal involves calculus, so it was a bit difficult to do using analog circuitry. In fact, VU meters were designed to yield an approximate RMS value of the incoming analog signals. But RMS calculations are easy in digital systems.

Figure 5.1
*An analog VU
meter.*

RMS stands for *root-mean-square*. The RMS value of a collection of numbers is found by dividing the sum of the squares of the numbers by the size of the collection and then taking the square root. For example, the RMS value of the set {1, 2, 4} is

$$\sqrt{\frac{1^2 + 2^2 + 4^2}{3}} = 2.65$$

Notice that this is slightly larger than the arithmetic average of the numbers, which is

$$\frac{1+2+4}{3} = 2.33$$

RMS meters work by computing the RMS value of a sliding-window groups of audio samples. The audio to be metered is *windowed*, or segmented, by the algorithm to update the meter several times per second. If the audio were not windowed, the meter would simply be displaying a running tally of the RMS value of the entire sound file up to that point, which is not as useful in most circumstances.

RMS metering is used in many areas besides audio, but it is particularly useful for our purposes because it yields a kind of average *power*, not the average amplitude, of a signal, which better describes how our ears perceive loudness. We'll return to this in Chapter 7, because this fact becomes really important when dealing with surround audio.

Peak meters, on the other hand, display the exact peak level of an audio signal. They can report instantaneous peaks, so that a single clipped sample of digital audio registers as an "overload" on the peak meter. Many peak meters on DAWs allow a peak-level hold display, wherein the peak level is "held" for a certain period of time, allowing peak levels to be read with greater ease. Peak meters are especially useful when mastering digital audio.

The rapid growth of multichannel/surround mixing has led to the creation of multichannel metering tools. Multichannel audio can be metered in several ways; perhaps the most common display is a collection of bar graphs, with each bar graph corresponding to a different audio channel. For example, a six-graph meter might display the peak or RMS levels of the front left, front center, front right, left surround, right surround, and low-frequency effects (LFE) channels. In fact, most surround-capable multitrack editor/sequencers feature displays like this for each surround audio track in a mix. (See Figure 5.2.)

Figure 5.2
Most surround-capable multitrack software programs provide multichannel metering capabilities. Shown here is a screen shot from Pro Tools; note the 5.1-channel meter on the output track at far right. © 2004 Digidesign, a division of Avid Technology, Inc. All rights reserved. Reprinted with permission of Digidesign.

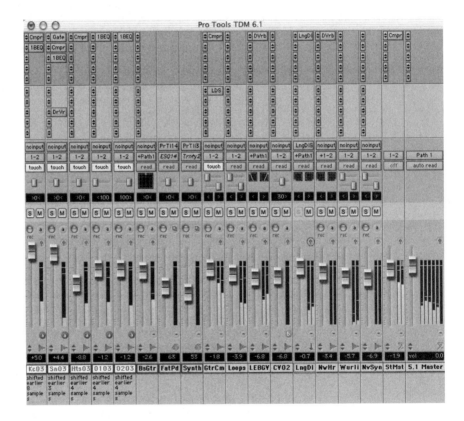

Frequency-Domain Analysis

It is often useful to visually examine the frequency content of audio signals. Frequency-domain analyzers perform a mathematical transformation of the input samples known as the Fast-Fourier Transform (FFT), and the FFT in turn tells us which frequencies are present in the signal and how "strong" each frequency is, among other things.

Frequency-domain analyzers are particularly useful when troubleshooting the acoustics of a listening environment. For example, if we play white noise (which contains all frequencies we can hear in equal amounts) into a project studio and record the sounding result into a DAW, we can discover where the *eigenmodes* (or simply "modes") of the room occur. Because the walls, ceiling, and floor in the room reflect sound somewhat and they are a fixed distance apart, the room resonates with some frequencies more than others. A well-designed studio is constructed so that these modes are distributed across all frequencies, so they tend to be less perceptible, but if you do not have the luxury of a custom-designed sound studio, chances are that clearly audible modes are present in the room.

Using a metering plug-in like SpectraFoo (Figure 5.3) from Metric Halo (http://www.mhlabs.com), we can examine the resonances that the room favors, which can help us to "tune" the room using equalization on the monitor loudspeakers. For example, if the room tends to rumble or resonate around 100 Hz, we might connect a graphical equalizer to the control-room outputs of our mixing console and "turn down" 100 Hz just a bit. By continuously adjusting the equalizer while monitoring the effect it has on the room's acoustics in a frequency-domain analyzer, the room can be properly tuned using the benefit of both our ears and our eyes.

Figure 5.3
Screenshot of Metric Halo's SpectraFoo metering plug-in.

Spatial Imaging Analysis

Spatial imaging analyzers have been around for awhile, but they are becoming particularly popular with the widespread growth of multichannel surround recording and mixing. Although they are generally a subset of time-domain analyzers because most of them work strictly in the time domain, we discuss them separately here because they are becoming almost as standard to DAW mixing applications as traditional time-domain metering methods.

Spatial imaging analyzers simply provide some kind of graphical display of the sound-field image in a mix (whether the mix is in stereo or surround) that can help mix engineers visualize and monitor several key spatial features in a mix:

- Is the soundfield lopsided? For example, is the front left channel producing substantially more audio material than the front right?
- Is the soundfield too wide or too narrow? If the spatial imaging analyzer indicates that most acoustical energy is coming from the front center, the mix may sound too monaural. If the analyzer indicates all sound is produced at the front left and right extremes with nothing in the center, the mix may sound too wide and one-dimensional.
- Are phase problems in the mix affecting the intended soundfield?

These three examples are clearly features that our ears can hear, but some mix and mastering engineers find it often useful to double-check their ears with their eyes to make sure they agree.

A popular spatial auditory display is loosely rooted in the concept of Lissajous figures. Lissajous figures, named after French mathematician Jules Antoine Lissajous (1822–1880), represent vibrational motion on a two-dimensional plot. A kind of mechanical predecessor to the modern electronic oscilloscope created these figures optically, and in fact Herman Helmholtz, the "father" of musical acoustics, used them to study tuning and musical scales and to measure the frequencies of sine waves. By using a sine wave of a known frequency to control the machine's horizontal input, a sine wave of an unknown frequency could be measured by mapping it to the machine's vertical control. Based on the frequency ratio between the known and unknown signals, one of several known patterns would result, as shown in Figure 5.4.

The study of various mathematical, dynamic, physical, and musical systems and their corresponding Lissajous figures they could produce spawned a great deal of esoteric research and a good number of books in the nineteenth century. Curiously, they have found new life in serving as the impetus for spatial imaging displays on the DAW.

Plug-Ins

Figure 5.4
Lissajous patterns.
(a) x and y are
both 440 Hz sine
waves (a) 1:1
frequency ratio) in
phase; (b) 1:1
frequency ratio
90° out of phase;

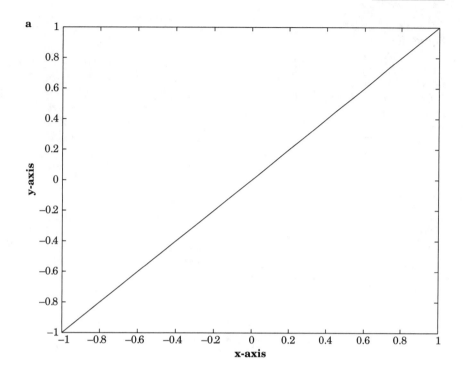

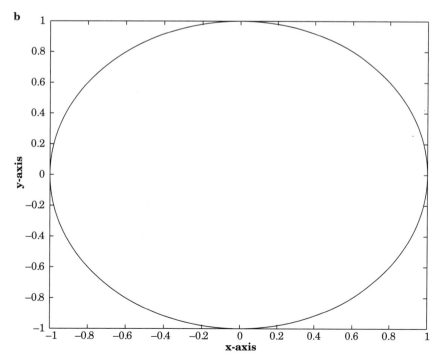

Figure 5.4

Lissajous patterns.
(c) 2:1 frequency
ratio in phase; (d)
2:1 frequency
ratio 45° out of
phase;

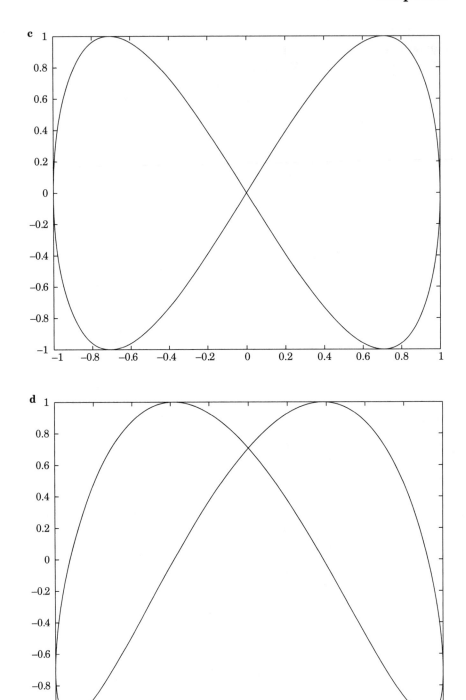

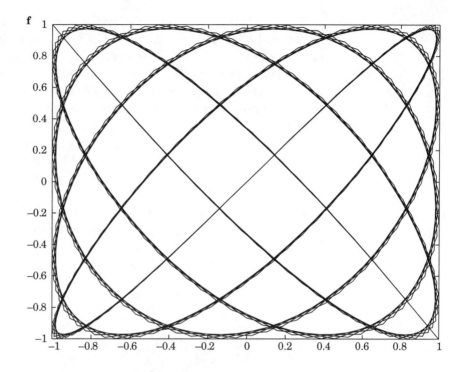

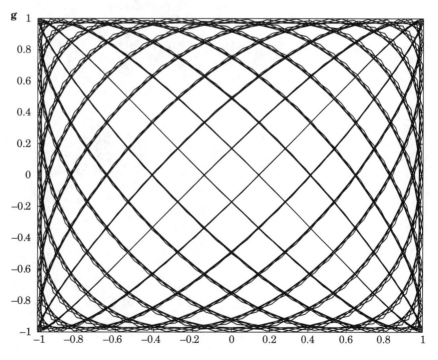

Several plus-ins use a display form loosely related to Lissajous figures called *polar patterns* as a kind of spatial imaging display. (The same kind of display is used to specify the directional response characteristics of microphones, as we will see in Chapter 8.) One simple way they can be employed is by mapping the instantaneous ratio of the left channel's amplitude envelope to the right channel's amplitude envelope to control a one-dimensional "puck." When the puck lies at the extreme left or right, this indicates a potentially "lopsided" spatial mix. This idea can be easily extended to display the spatial content of surround mixes. DAW plug-ins like Digidesign's SurroundScope (Figure 5.5) and Metric Halo's SpectraFoo provide visual feedback to the spatial content and sound field of a mix using similar techniques. A perfect circle, for example, would indicate that the mix is evenly distributed among all loudspeakers in a surround mix; the example in the forefront of Figure 5.5, however, would indicate a right-heavy mix.

Figure 5.5
Screenshot of Digidesign's SurroundScope plug-in. © 2004 Digidesign, a division of Avid Technology, Inc. All rights reserved. Reprinted with permission of Digidesign.

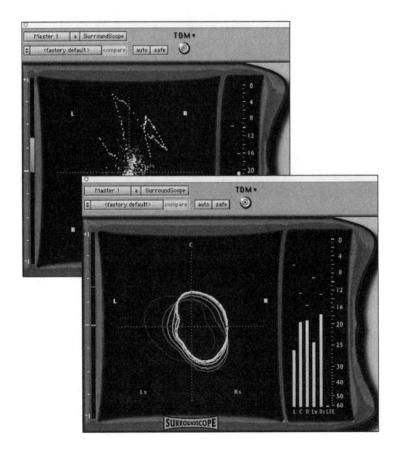

Comparable standalone hardware imaging tools for spatial analysis, particularly for surround audio analysis, can cost well over US$10,000. The high quality and low cost of the many software spatial analyzers currently available bring yet another added dimension of value to the DAW.

Tempo Trackers

Musicians from many different walks of life need to be aware of the tempo of a musical passage: DJs, who frequently cross-fade similar-tempo music; recording and mix engineers, who must tempo-lock different tracks; and sound designers and film scorers, who may find it useful to query a large database of sound files based on tempo.

Tempo detection is generally quite easy for humans to perform. When we hear grooves, recurring patterns of sound, or beat-oriented music, many of us naturally start tapping our feet in time to the music. The number of times we tap our feet per minute represents the tempo in beats per minute (BPM) of the music, just as we can measure our pulse rate by counting the number of times our heart beat per minute. Although this process is easy for us to do and we come by it quite naturally, it is a little more difficult for computers to perform accurately.

The problem can be broken down into two components: beat detection and tempo induction. Simply stated, the first part, *beat detection*, is the hard part, and the second part, *tempo induction*, is the easy (well, easier, in any case) part. The psychoacoustic processes that enable our brains to induce beats from music are quite complicated and not fully understood yet, so it is currently somewhat difficult to write a computer program to do this automatically with an extremely high degree of accuracy. The following "problems" can make the automatic detection of beats from sound files difficult:

- The strongest beat in a measure of music, usually called the *downbeat*, is not necessarily any larger in amplitude than other beats.
- Syncopations and rests can mean that either no change in sound or complete silence occurs directly on a beat.
- Often, humans do not agree on what the basic beat quantum in a musical passage is (e.g., quarter notes, half notes, eighth notes, etc.)
- The tempo can often change within a sound file, sometimes quite drastically.

Once we have induced beat locations and their inter-onset intervals (IOIs) (see Figure 5.6), we can then count how many of them occur in any given minute to find the tempo in BPM.

Figure 5.6

Inter-onset intervals between beats.

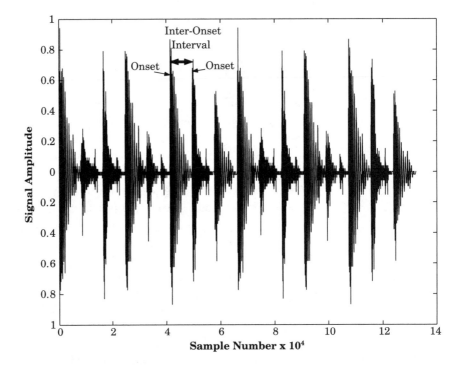

Many different algorithms for computing tempo automatically currently exist. Some of them work primarily on music with drum beats, while others perform well on almost any kind of music. And even as humans cannot always unilaterally agree on what the tempo is, it should be noted that no one automated guess is ever completely correct. And just as we often don't agree on the basic beat quantum (half, quarter, eighth, etc.), beat detection algorithms often yield so-called "tempo octave" errors, or answers that are "incorrect" by a 1:2 or 2:1 ratio.

Some multitrack editor/sequencers include built-in beat-detection functions. For example, some Pro Tools software features Beat Detective, a robust and configurable tool. Technically, Beat Detective is not a plug-in, as it is included standard on Pro Tools TDM software, but other beat-tracking plug-ins are available, and more are sure to come.

Pitch Trackers

Pitch trackers also emulate aspects of the way humans hear sound by trying to mimic the way we perceive pitch. Recall that pitch is not a physically measurable and precise quantity; it is the psychoacoustic correlate of frequency. Frequencies are easily measured using mathematical tools like the FFT, but pitch is the result of the human brain's processing of various frequency information. For example, it is quite easy to trick our ears into hearing one pitch when we're actually playing another. We'll come back to this topic in a bit when we return to plug-ins that deal with psychoacoustic enhancement of audio.

Simply stated, a pitch tracker performs an algorithm on a selection of audio and reports the computed pitch. This can be useful when trying to lay down accompaniment in post-production for a previously recorded singer. Pitch trackers are most useful, however, when they are used not as a simple auditory display, but rather as one component of a pitch-correction algorithm that tries to "fix" the pitch imprecision of a recorded singer or instrumentalist. Pitch trackers can also be used creatively to map a melodic line from one audio file to another. For example, we could extract the pitch envelope from a recording of a baby's crying and apply it to another recording of a dog's barking.

Experimental

Various experimental auditory display plug-ins are also available for the DAW, and they are more often used in the computer music research community than in commercial mixing applications. We can generate state-space (also known as *lag space*) and wavelet transform plots, for example. Interesting results have been obtained by researchers in characterizing timbral features of different sounds using experimental auditory displays like these. For example, consider the audio state-space diagram, which is a two-dimensional plot of the samples of an audio signal versus a delayed version of the same samples. Diagrams like these can visually convey sonic information that may, with practice and coupled with a good set of ears, provide us with another layer of useful information about the tonal qualities of a mix.

Effects Processors

Effects processors are probably what most people first think of when they hear about plug-ins, and they are the greatest in number. Not only does digital signal processing allow any conceivable audio effect to be produced, but, owing to the processing power of modern computers, many DAW plug-in effects processors far surpass processors previously only available as standalone hardware devices in terms of cost, value, bit depth, sampling rate, and sound quality.

Effects processors can be broken down into twelve functional groups: dither, reverberation, delay and echo, spatial processing, chorusing and flanging, equalization, companding, psychoacoustic processors, pitch shifting and time scaling, pitch correction, gates and triggers, and audio signal restoration. Another family of miscellaneous and experimental effects processing plug-ins are also described, followed by a brief discussion of concepts behind multieffects chaining.

Dither

Dither, a middle-English term meaning "tremble," is simply low-level noise that is added to a digital audio signal. Why would we want to do that? The ironic result is that the audio sounds "better" with some low-level noise added to it, and here is why.

Try this experiment: listen to the tail end of various compact disc tracks, and as the track ends, turn up the volume on your stereo to try to keep the music at the same loudness. Eventually, you probably start to hear some kind of "digital fuzz" where individual quantization levels are being turned on and off rapidly. As the audio signal level gets lower, the quantization noise gets higher. That's because the ratio of the amplitude of the signal to the size of the nearest quantization level is getting bigger, so the quantization noise is becoming larger relative to the signal's amplitude.

As the signal approaches zero in amplitude, there are times when the analog signal was rounded up to the first quantization level and times when it was rounded down to zero. This process happens alternately as the signal fades away. This "bit flipping" creates a little bit (no pun intended) of audible noise, and it results from the fact that we're trying to ask a digital system to produce less than the lowest quantization level. Said another way, we are asking it to encode less than a single bit's worth of information. (See Figure 5.7.)

Plug-Ins

Figure 5.7
Using dither lets
less than one bit
of information be
encoded. (a)
Original analog
signal; (b)
quantization of
undithered signal
with original
analog signal
overlaid;

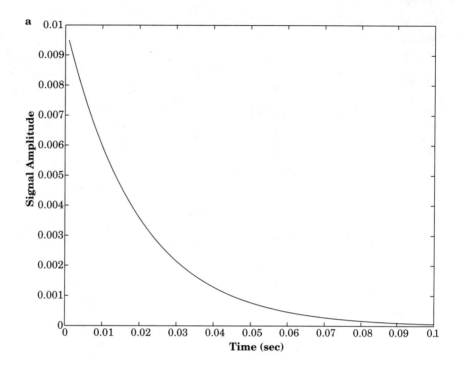

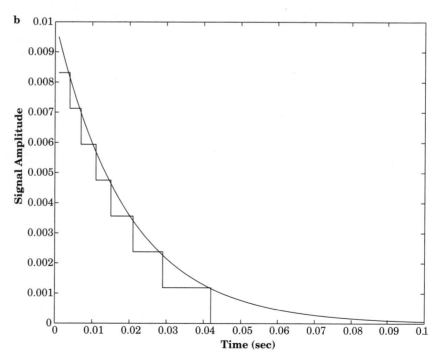

Figure 5.7
Using dither lets
less than one bit
of information be
encoded. (c)
quantization of
dithered analog
signal; (d)
comparison of
undithered and
dithered signals.

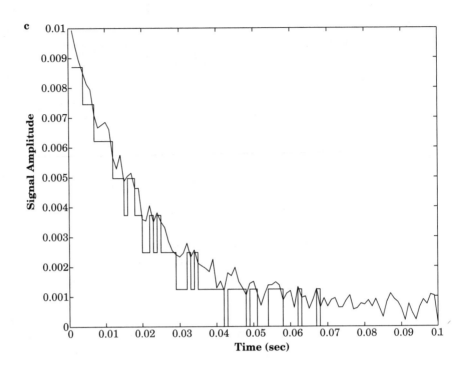

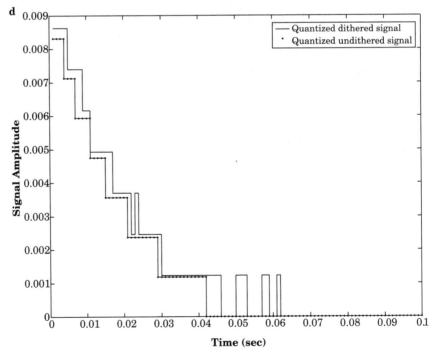

And here is where dither comes to the rescue. By simply adding less than a single bit's worth of noise, the dithering noise signal can carry enough "oomph" to make our D/A converter round to the nearest quantization level more effectively and more accurately. The analogy usually used to help us understand dither in physical terms is this. Pretend the words on this page represent analog audio information. Place your hand, fingers slightly spread apart, directly in front of your face, and close one of your eyes. It's now quite difficult to read the words, but if you rapidly move your hand (sort of like introducing dither noise), it becomes much easier to read the words. The analogy isn't perfect, but it should make intuitive sense.

When is dithering needed? Whenever we need to encode less than one bit's worth of information, which is the case whenever we reduce one word length to another. For example, many DAW software programs perform operations internally at 32 bits, because performing mathematical operations (remember digital signal processing?) on 16-bit samples, for example, can quickly result in 32-bit words. (Think of a simple example: multiplying 16×16, [i.e., two 4-bit words], yields 256, which requires 8 bits to express.) So dithering is often the final stage in mastering a mix for production, because the higher-resolution internal word length of our DAW software will probably need to be bit-reduced for storage on a compact disc, DVD, SACD, or the Internet.

Many different dithering algorithms exist, and some sound better than others. An interesting DAW dithering plug-in technology is Increased Digital Resolution (IDR) from Waves (http://www.waves.com), which applies less dither noise to audio signals at lower amplitude levels. This makes sense, because the noise floor will be correspondingly reduced at lower audio levels, helping to keep the overall signal-to-error ratio more or less constant. Remember, adding dither noise is a tradeoff: We're trading the audio distortion that results from quantization error for a higher noise floor. But this trade-off tends to work in our favor; just as we can discern individual voices in a crowd of people, we can pick out signals less than one quantization level in the presence of dither noise.

The whole story of dither can get fairly complicated quickly, and we could use lots of hairy equations to explain it in greater detail. Or, you can just remember this: Adding a small amount of dither noise to a digital signal can actually make it sound better!

Delay and Echo

Two of the simplest and yet often musically effective effects plug-ins are delay and echo. Virtually all two-track and multitrack editors include these effects right out of the box. And using some of the audio signal processing techniques that we discussed earlier in this chapter, we can explore them in a bit more technical detail than might otherwise be possible.

First of all, what is the difference between delay and echo? (Remember that delaying a signal is the fundamental building block of virtually all signal-processing algorithms.) Some plug-ins do not differentiate the two terms, but in the strictest sense, delay plug-ins simply pass a single delayed version of the input audio signal, optionally adding the original input as well. This this can can sound sound like like this this. Of which family of digital filters does this remind you? This by itself is a simple feedforward filter:

$$y_t = x_t + Ax_{t\text{-}delay}$$

This filter says simply, "the output of the filter is the input to the filter added to an attenuated, delayed copy of the input to the filter." Again, y_t is the tth output sample, x_t is the tth input sample, A is the attenuation amount of the delayed version of the input, and *delay* is the number of samples to delay the audio. If our sampling rate is 44.1 kHz, then setting delay = 44,100 would correspond to a 1-second delay. That is, the desired delay expressed in samples in simply the desired delay time in seconds multiplied by the sampling rate. Virtually all delay plug-ins allow the user to specify the delay in terms of seconds, not in the number of samples; they perform this simple computation in the background because it's usually more useful musically to think in terms of delay time rather than delay samples.

Echo, on the other hand, is not made using a feedforward filter; it is made with a feedback filter. This is because echo plug-ins delay not only the input to the filter, but the output of the filter as well. Psychoacoustically, the delay time must typically be at least around 50 or 60 milliseconds for us to perceive discrete events. The sonic result is that we hear multiple repetitions of the delay, each more attenuated than the other. Think of yourself yelling "echo" into the Grand Canyon: You would hear "echo" multiple times, each one a little softer than the previous. I'll leave it to you as an exercise at the end of this chapter to ponder what the feedback filter equation might look like for a simple echo. Note that many plug-ins encapsulate both delay and echo functionality within one piece of software.

Delay and echo can be effectively used in different musical scenarios. They are often applied to the lead vocal line in popular music, often with the delay times tempo-locked to the underlying groove, to accentuate it and for emphasis. Echo can also be used to create simple loops and grooves from a single measure of audio: Just set the feedback time to the length of the measure, set the feedback amount to 100 percent, and you're looping!

Echo also serves well as a virtual arpeggiator. By playing a simple melody, like that shown in Figure 5.8a, the result after application of an echo plug-in might be slightly more interesting and complex, as shown in Figure 5.8b.

Delay and echo can serve on their own to create different sound spaces and sonic environments, and they themselves are the fundamental building blocks of the next plug-in algorithm we discuss, reverberation.

Reverberation

Consider two different scenarios: popping a balloon in your bathroom, and popping the same balloon in a large cathedral. What makes these two events sound so different? Clearly, the acoustics of the rooms explain the difference. From practical experience, you know that the sound will die away quickly in a small room like a bathroom, while it will linger around a while longer in a large space like a cathedral.

Reverberation occurs whenever sound is produced anywhere near a collection of surfaces that can reflect sound waves, causing a "wash" of the resulting reflections to reach our ears very close together in time. (Remember, if the reflections are far enough apart so that we can discern discrete instances of the source but with no trailing "wash," we only have echo, not reverb.)

Reverberation ("reverb" for short) is composed of three distinct components that respond to the initial sound source, in our case a balloon pop. (See Figure 5.9.)

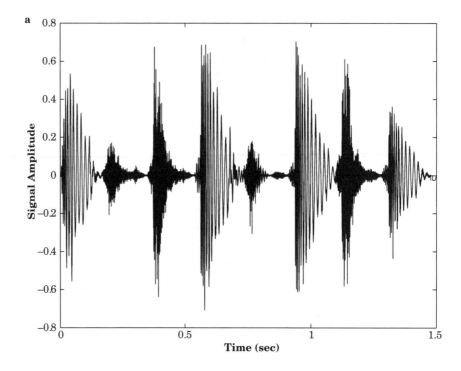

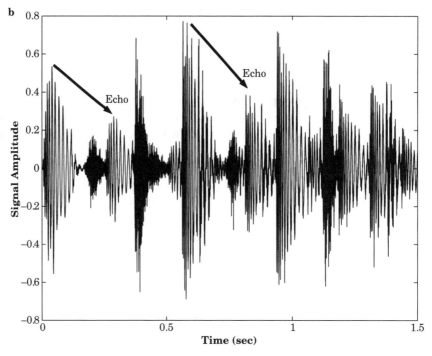

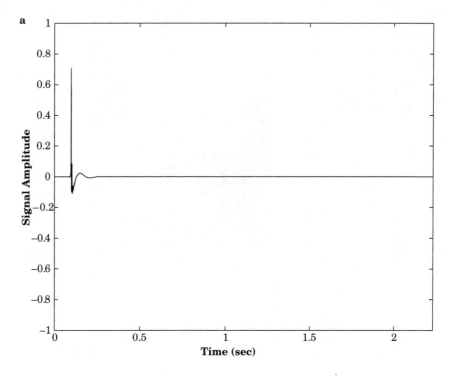

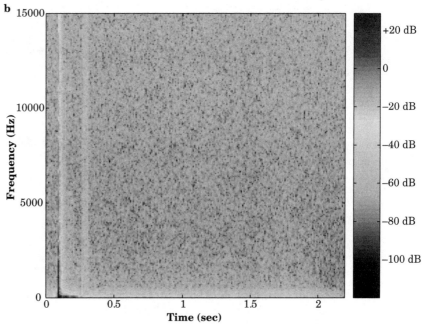

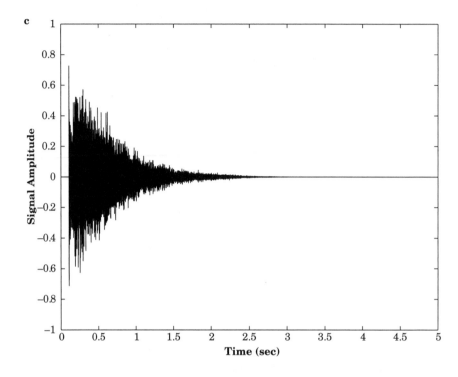

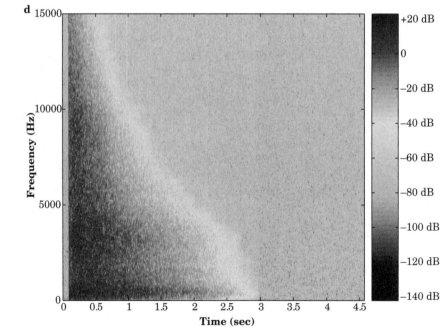

The first element is called *direct sound*, which is a direct result of the finite speed of sound in air. Even without reflecting surfaces, a certain amount of time would have to pass from the popping of the balloon until we actually heard it pop. The first balloon pop we perceive is the direct sound.

The next portion of reverb is the series of *early reflections* that then occur. These are a collection of echoes of the impulse sound's reflecting off various surfaces in the acoustic space.

Finally, the *reverberant sound*, which I've been referring to as a "wash" of sound, occurs. This is the collection of reflections that occur too close together in time to be perceived as independent sonic events. This is also typically the part of artificial reverberation algorithms that separates the chaff from the wheat. It's easy enough to artificially generate direct sound and early reflections, but the reverberant sound is considerably more involved. Collectively, the early reflections and reverberant sound portions of a reverb are called *indirect sound*.

One other term worth defining in our discussion is the *reverberation time*, sometimes called the RT60 level. This is the amount of time it takes for the reverberant sound to fall 60 dB lower than, or 1/1000 the value of, the direct sound. Note that the specific aspects of the components of reverberation—the time delay between direct sound and the start of early reflections, the ratio of direct sound to indirect sound, the reverb time, and so on—greatly affect our perception of the space in which the impulse occurs.

Algorithms for artificial reverberation fall into two distinct categories: *time-domain algorithms* and *convolution-based algorithms*. We talk about each of them briefly here, because reverb plug-ins of both types are in common use. Be wary of using too many reverb plug-ins, though: They can really tax your DAW's processor! Some are less taxing than others, but a good reverberation, particularly when used in a surround mix, can consume a considerable portion of the available CPU speed (whether on your host computer or a separate DSP accelerator card).

In 1962, Manfred Schroeder wrote a classic article in which he proposed a digital reverberation algorithm. Now called the Schroeder Reverb, it forms the basis for many of the time-domain algorithms that have followed since. In simple terms, the time-domain algorithms based on this classic design employ a large bank of feedback filters (particularly comb filters, and allpass filters, which I leave it to you as an exercise to learn about). These feedback filters are strategically designed in a complex network to mimic the acoustical feedback that occurs naturally in reverberant environments. Time-domain reverberation plug-ins, such as Lexicon's LexiVerb plug-in (http://www.lexiconpro.com), use high-quality reverb algorithms that allow a plethora of user settings to create different acoustical spaces.

The second family of reverberation algorithms is based on a signal-processing trick called *convolution*. Convolution can be thought of in several ways, but the simplest is as a mathematical operator, like addition or multiplication. In fact, it has been given its own symbol in signal processing, the asterisk ("*"). Rather than first explaining convolution in detail with words, I offer a numerical example for you to figure out. We will convolve a with b, shown below.

$a = \{1, 2, 3\}$
$b = \{4, 5, 6\}$

```
              1      2      3
6      5      4
              (4)

              1      2      3
       6      5      4
              5  +   8 = (13)

              1      2      3
       6      5      4
              6  +  10  +  12 = (28)

              1      2      3
                     6      5      4
                    12  +  15 = (27)

              1      2      3
                            6      5      4
                            (18)
```

$a * b = \{4, 13, 28, 27, 18\}$

Here is how we arrived at the result. Convolution is simply a process of folding, shifting, and integrating. First, we reversed the order of the elements in b. Next, we shift b underneath a until they have one element lined up on top of each other and multiply the common elements together, yielding $1 \times 4 = 4$. Next, we shift b to the right again until two elements are in common. Integrating the common elements, we get $(1 \times 5) + (2 \times 4) = 5 + 8 = 13$. So 13 is now the second element of our resulting convolution. This process is repeated until no more elements are lined up in common between a and b.

Okay, now you can wipe the drool off your face. Consider an interesting example, in which the first set to be convolved (a in the previous example) is the collection of audio samples from a recording of an ideal balloon pop:

$$\text{Balloon pop} = \{1, 0, 0, 0, 0\}$$

Now, consider the second set (b in the previous example) is the recorded *impulse response* of a room:

$$\text{Impulse response} = \{1, 0.5, 0.25, 0.125\}$$

The impulse response of an acoustic space simply describes how the room responds to an acoustic impulse, like our ideal balloon pop or an acoustic firing pistol. Notice how the given impulse response decays over time: Each subsequent sample value is half the value of the previous.

Here's where we are headed with this: Using convolution, the impulse response of a room can be applied to any digital signal, not just the original balloon pop, to make it sound as though the convolved signal were produced in the room in which the impulse response was recorded. I will leave it as an exercise for you to figure out the convolution of our ideal balloon pop with the given impulse response.

This becomes musically useful when convolving a sound file with the acoustic impulse response of a real acoustic environment, like a bathroom or a cathedral. And this is just how convolution-based reverbs, often called *sampling reverbs*, work. Because the number of samples to be convolved can be very high (even at only 44,100 per second), convolution-based reverbs can be especially computationally taxing, but a couple of mathematical tricks can speed things up a bit (which I'll leave for another day).

Although the technique of convolution has been known for many years, computers were not fast enough to perform convolution of audio signals in real time until several years ago. But now, several great plug-ins based on this principle are available, including Altiverb from AudioEase (http://www.audioease.com). Many audio files of impulse responses from real acoustical environments around the world are available all over the Internet.

To my ears, most convolution-based artificial reverbs sound more natural than their time-domain counterparts. The tradeoff is usually that time-domain reverb plug-ins typically offer greater control over their parameters and may demand less signal-processing horsepower.

Musically, when is a reverb plug-in useful? I offer three primary reasons.

- Reverberation can impart a sense of "realness" to an abstract sound object. Because reverberation only occurs when a sound occurs within a physical space, sounds placed in a multitrack mixer within a reverberant field can convince us of their reality.
- Conversely, because a reverb plug-in is an artificially generated digital audio effect, it can be altered. For example, the physical spaces of any dimension can be evoked; or perhaps in conjunction with a time-domain reversal of a reverberated sound, the order of occurrence of direct sound and indirect sound can be flipped, eliciting a sense of the physically impossible.
- Reverberation can blend together sound objects. It can also "smooth over" imperfections, but as is often said, reverb is like sugar! A little tastes great, but too much can impart a saccharine quality to a mix.

Because using a reverb plug-in typically has the physical interpretation of placing a sound within a real room or space, it calls for caution. If you process one audio track with one setting of a reverb plug-in while processing another track in the same mix with a different setting of the reverb, that may sound like each track was recorded in a different room. (This may or may not be want you want.) Likewise, consider the effect of applying a reverb plug-in to a sound that was itself recorded in a physically reverberant space, which has the physical interpretation of placing one room inside another room.

Chorusing and Flanging

Two related time-domain plug-in effects are *chorusing* and *flanging*, both of which can serve to thicken the texture of an audio track. They both work by mixing delayed copies of an audio signal with itself and are very computationally simple; hence they are not processor hogs. The texture is thickened with both effects not only because delayed copies of the audio signal are mixed with the original, but the delay times vary in time as well.

Flanging occurs when a signal is delayed for a very short time (typically less than 20 milliseconds) and mixed with itself. The characteristic flanging effect occurs when the delay time of the feedback is varied, usually sinusoidally; that is, the delay time may vary between 5 and 10 milliseconds at a rate of 3 Hz, or three full cycles per second. This leads to phase cancellations and reinforcements, leading to a "phasiness" quality, although the resulting sound usually is perceived to emanate from a single source.

Chorusing, on the other hand, involves the mixing of multiple delayed copies of a signal with itself. The delay time used by chorusing plug-ins is also longer, on the order of about 20 to 50 milliseconds. Although these are still very short delay times, they are long enough for our ears to perceive that multiple sound sources are present in the signal, like a chorus of singers. The delay times corresponding to these different effects are summarized in Table 5.1.

TABLE 5.1

Approximate Delay Times of Flanging, Chorusing, and Echo

Flanging	Chorusing	Echo
0–20 milliseconds	20–50 milliseconds	50+ milliseconds

Equalization

Equalizers are frequency-selective amplifiers: They can provide gain to selected frequencies of an audio signal while attenuating the level of other frequencies. Equalizers can be characterized in two broad ways: based on the frequency band of interest (e.g., bass, mid, or treble) in which they operate and what they do to that band; and also according to the way in which they are controlled by the user. We will talk about each of these in turn.

Four basic categories of equalizers are available based on their region of operation as we discussed earlier: low-pass, band-reject, band-pass, and high-pass. As the names suggest, low-pass filters attempt to let low frequencies through while attenuating high frequencies. (Think of this as turning up the "bass" knob in your car stereo.) Band-reject (also called *notch*) and band-pass filters selectively cut or pass only a specified frequency band. For example, a band-pass filter might be configured to only pass audio information in the 500 to 700 Hz range, or we might use a 60-Hz notch filter to remove ground-loop interference that has crept into our DAW recording chain. Finally, high-pass filters are the converse of low-pass filters: They pass high frequencies will attenuating low frequencies.

Two basic control configurations are possible: parametric and graphical, although some equalizer plug-ins incorporate aspects of both types. Parametric equalizers provide controls over the precise frequency you wish to change (called the *center frequency*), the amount of gain or attenuation to apply to that frequency, and the selectivity of the filter to the surrounding frequencies on either side of the center. They are called parametric because they allow parametric control over the filter's shape, and a true parametric equalizer might contain many such independently controllable filters. The frequency selectivity, or resonance, determines whether the filter's shape is narrow or more broad. It can be specified in terms of the bandwidth (that is, the frequency range where the response of the filter is within 3 dB, or 50 percent, of the peak level). This is shown graphically in Figure 5.10.

The amount of a filter's resonance can also be specified in terms of Q on most parametric equalizer plug-ins, which is simply the ratio of the center frequency to the bandwidth. A small Q value, like 0.1, indicates a broad filter characteristic, while a large value, like 10, indicates a much narrower resonance. Many interesting effects can be achieved by varying the Q of a filter over time; it's almost like zooming in and out with a camera, because you can "zoom in" on a specific frequency by raising the Q.

A graphical equalizer, on the other hand, provides a single attenuation or gain slider or knob for a fixed set of frequencies. A one-half-octave graphical equalizer, for example, provides two separate attenuation/gain controls per octave; a one-third-octave graphical equalizer provides three controls per octave. Unlike a parametric equalizer, the center frequency and bandwidth are not controllable. The International Standards Organization (ISO) has standardized the center frequencies for various graphical equalizers based on how many frequency controls they provide per octave.

Figure 5.10
(a) The bandwidth
and center
frequency of a
resonant
parametric filter;
(b) magnitude
response of high-
Q resonant filter;

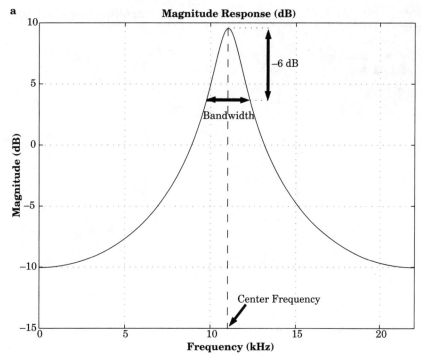

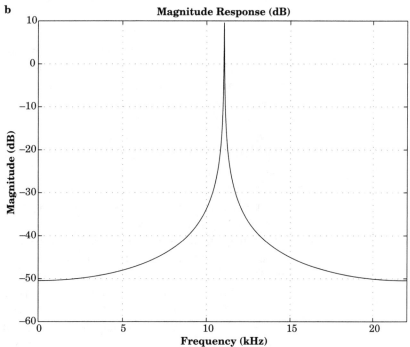

Figure 5.10
(a) The bandwidth and center frequency of a resonant parametric filter; (b) magnitude response of high-Q resonant filter;

Figure 5.10
(c) magnitude
response of a low-
Q resonant filter.

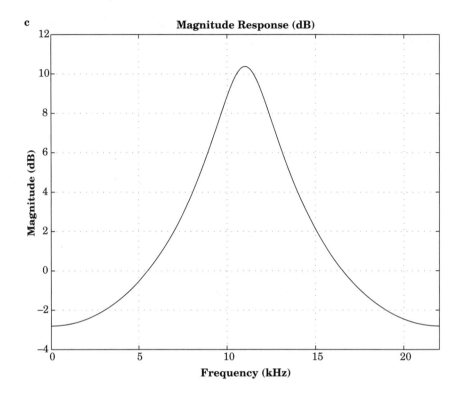

c

Figure 5.10
(c) magnitude response of a low-Q resonant filter.

Note that relatively simple equalizers are typically found on the channel strip of hardware mixing consoles. However, to minimize CPU load, most multitracks' strips do not equalize by default. But they are easily invoked with an equalizer plug-in. Most multitrack software includes one or more such plug-ins that can of course be applied to audio, auxiliary, or master tracks.

Equalizers can be used in three main applications:

- **"Let's get funky."** Equalization can be used creatively on individual tracks to exaggerate or emphasize a frequency range, coloring and possibly diminishing the recognizability of the source.
- **"Putting lipstick on the pig."** Equalization can, to a limited extent, compensate for certain defects in a recording. But a pig is still a pig! It can also be used to compensate for the nonlinear nature of our hearing, as we discuss later in the chapter.
- **"Turn off that noise!"** A quick-and-dirty way to minimize hiss is to roll-off the high-end frequencies. A side effect is a little loss in presence; the sound can seem farther away, and particularly for

voices, can diminish intelligibility. It's better to use a noise-reduction plug-in, but an equalizer may work in a pinch.

- **"Spread yourself out."** A stereo panpot only provides control over the left-to-right soundfield. Low-pass filtering with an equalizer plug-in, perhaps coupled with a reverb, can "move" a sound farther away from the listener. (Notice how the farther away someone speaks to you, the darker their voice can seem.)
- **"Putting on the finishing touches."** Once a mix is almost complete, equalization is most always performed on the master track to adjust the overall tonal balance, for example by making the mix darker or brighter.

Waveshaping and Distortion

Waveshaping is a general family of techniques that simply distort the time-domain waveform of a signal. By defining an arbitrary transfer function shape (remember, transfer functions define the ratio between the output and the input of a process), the input signal to a waveshaping plug-in can be warped based on its current amplitude. For example, low-level and high-level amplitudes might be passed without change, while mid-level amplitudes might be greatly amplified. In general, this can greatly impact the resulting spectrum of the output sound, and it can do so at very low computational cost.

When we say "waveshaping," what we are really talking about is *distorting* the input signal. Distortion, as a musical term, evokes its own guitar-centric connotations of grunge and death metal, perhaps, but it is simply waveshaping. Distortion plug-ins impart rich harmonic content to sounds by squashing the high-amplitude portions down a bit, just as an amplifier driven too "hot" will do.

Dynamics

A dynamics processor plug-in is an essential tool to own for a DAW user. Dynamics processors simply expand or compress the dynamic range of a signal. A dynamics processor that expands the dynamic range is called an *expander*, and a processor that compresses the range is called a *compressor*. Collectively, these are called *dynamics processors* or *companders*, and they are used all the time on audio tracks, auxiliary tracks, and master tracks in a mix. A good dynamics processor plug-in is especially crucial in the final stages of a mix-down.

A compressor plug-in effectively "squashes" a sound: It makes the loud parts sound softer and the quiet parts sound louder. A negative side effect of compression, however, is that it raises the noise floor. That is, because the signal-to-noise (and the signal-to-quantization-error ratio as well) is lower at smaller amplitudes in a signal, the compressor will raise the level of this noise.

Compressors usually include four primary controls used to define their transfer functions: *gain*, *threshold*, *ratio*, and *attack/release times*. (See Figure 5.11.) The gain control simply amplifies the signal before the compander kicks in. For example, if you are attempting to make a sound file sound louder, you can raise its total RMS amplitude level by turning up the gain a bit on a compressor.

The companding ratio defines the extent to which the compander compresses or expands the signal after the input signal has exceeded the specified amplitude threshold. The transfer function (that is, the ratio of output to input) of a compander is linear up to the defined threshold, after which the transfer function's slope is dictated by the companding ratio. (By the way, the point at which the compander kicks in is usually called the *knee*, and it may be a hard edge or a smooth transition.)

Consider a 5:1 compressor, for example. Once the input signal's amplitude exceeds the specified threshold, the output of the compressor only rises one-fifth as much as the input, hence squashing the dynamic range of the signal. The output of a 5:1 expander, on the other hand, rises five times as much as its input once the threshold has been exceeded.

The last primary setting of a dynamics processor are the attack and release times, usually specified in milliseconds. These quite simply determine, respectively, how quickly the companding ratio is applied or released once the input signal has made a corresponding transition across the threshold. Setting these numbers too low can make a compander sound very artificial, while setting them too high might not catch any transients.

Companders can take any shape, not just the linear ones shown in Figure 5.11. Furthermore, some dynamics processing plug-ins can be set to operate differently on different frequency bands. The MOTU MasterWorks Compressor, for example, is a three-band compander that allows user-definable frequency bands. To tighten the low-end and high-end in a mix, for example, while expanding the dynamic

Figure 5.11
Transfer function
of a compressor.
Note the gain,
knee, threshold,
and compression
ratio.

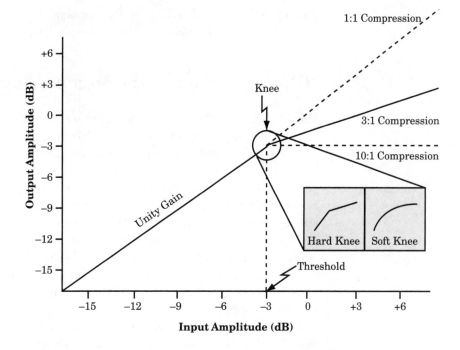

range of the midrange, you might compress only frequencies below 150 Hz and above 8 kHz while applying an expander to the middle frequencies.

Once a compander is set to a compression ratio of 10:1 or more, it is usually called a *limiter*, because it severely limits the maximum possible output amplitude. Conversely, once a compander is set to an expansion ratio of 1:10 or less, it is usually called a *gate*, because it "gates out" low-amplitude signals, like the underlying noise floor in quiet sections of a recording. When a gate is specifically used to remove noise during these quieter sections of a sound file, it is often called a *noise gate*. Noise gates are generally not as effective as spectral noise-reduction processing algorithms, however, partly because they do not remove any noise from signals whose amplitudes are above the specified threshold.

Triggers

A trigger follows the general contour of the amplitude envelope of a sound using an *envelope follower*. Once a target amplitude threshold has been reached, a trigger can send a control signal to...well...trigger another effect or send a MIDI message. These can be useful for making a sound file "play" a software synthesizer, for example.

"Classical" Computer Music Effects

Many "classical" effects processes that were used a great deal in the early analog days of electronic music have found new life on the DAW. These include amplitude modulation, frequency modulation, various cross-synthesis techniques, and many others.

With amplitude modulation (AM), the corresponding samples of two signals are simply multiplied together, resulting in the imposition of one's amplitude onto the other. (See Figure 5.11.) (A special case of amplitude modulation, called *ring modulation*, occurs when the modulating signal has only positive sample values, which you can do to a sound file by rectifying it in a two-track editor.) A musical analog of amplitude modulation in the physical world is tremolo.

Frequency modulation (FM), on the other hand, results when one signal (called the *modulator*) is used to dynamically control the playback frequency of another signal (called the *carrier*). A musical analog of FM in the physical world is vibrato. Consider a soprano who sings a note whose fundamental is centered on A = 440 Hz, but her vibrato causes the A to deviate periodically a total of 10 Hz on either side of 440 Hz, about five times per second. Here, we have a 440-Hz carrier, a 5-Hz modulator, and a index of modulation of 10/5 = 2. (The *index of modulation* is the amplitude-to-frequency ratio of the modulator; think of it as the "depth" or "amount" of modulation.) The result is that the soprano's fundamental moves between 430 Hz and 450 Hz five times per second (Figure 5.12).

Spatial Processing

Many spatial-processing plug-ins are available, and more are continually developed as surround mixing becomes more popular. Spatial processors fall into four categories:

- **Panpots.** Although all multitrack mixers include stereo panpot controls ("panners"), and many of them include surround panners, panpot plug-ins are available to provide other options and support

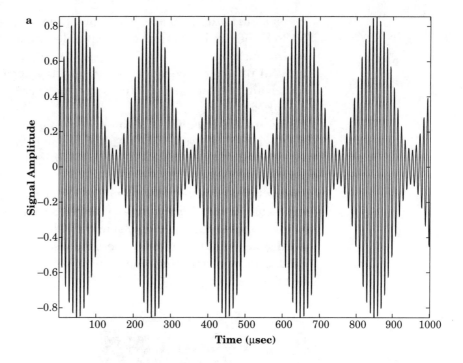

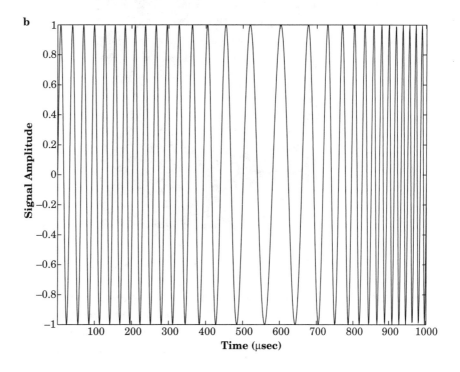

various surround formats. (One such plug-in is shown in Figure 5.13.) Some can be controlled in real time by external joysticks and control surfaces.

- **Spatial image processors.** With these, the apparent sound stage can be altered; for example, the perceived space can be widened or narrowed.
- **Distribution encoders.** If you are mixing for a distribution format like DVD-Video or DVD-Audio, you may wish to encode your mix in a supported format, like Dolby Digital or DTS, using an encoder plug-in.
- **Recording decoders.** As we will see later when we get to recording, some special microphone arrangements require decoding for proper representation of the sound field. These include the M-S arrangement and ambisonic microphones. Special plug-ins can be used to decode recordings made in these formats so that they can be mixed properly on a DAW.

Figure 5.13
Screenshot of AudioEase's Peter Pan spatial-processing plug-in.

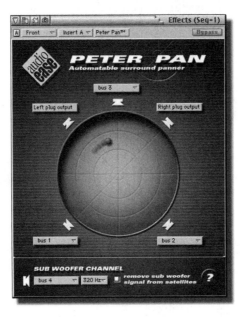

Psychoacoustic Processors

Psychoacoustic processing plug-ins attempt to exploit the characteristics of human audio cognition to achieve a desired effect. The idea is not new: Baroque pipe organs sometimes included a stop called a "Resultant" or "Resultant Bass." By pressing only one pedal to produce a bass tone, a resultant stop would activate the corresponding pipe as well as a pipe a perfect fifth higher. The sonic effect—the "resultant," you might say—is that we perceive one tone an octave lower. Resultants are still found on many pipe organs today.

MaxxBass (http://www.maxxbass.com), from Waves (http://www.waves.com), is a DAW plug-in that takes this idea to another level. Applying MaxxBass to an audio track performs a complex signal-processing algorithm that can cause listeners to perceive an octave subharmonic in the signal.

Figure 5.14
Screenshot of
Waves' MaxxBass
plug-in. © Waves
Audio Ltd.
www.waves.com

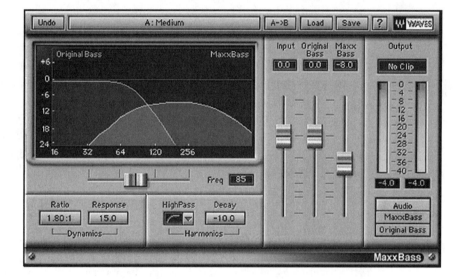

Human psychoacoustics provide many opportunities for us to exploit for musical gain. In fact, psychoacoustic processors in general form a major component of compression schemes like those in MP3 and MPEG. And as we continue to learn more about how we hear, surely many more psychoacoustic processor plug-ins will be developed.

Frequency Shifting and Time Scaling

The simplest way to change the frequency scale or time scale of a sound file is to play it back at a different sampling rate than that in which is was recorded or synthesized. However, doing so inextricably links the pitch-shifting ratio to the time-scaling ratio: If you read through a sound file faster, the frequencies will increase, but the time scale will be shorter. In a sense, the independent control of the frequency scale and the time scale of a sound is one of the pioneering (and in fact earliest) audio achievements of the DAW.

As an historical sidebar, before the DAW, an attempt was made to provide independent control over time and frequency scales using audio recorded onto magnetic tape. The Springer Machine was used throughout the 1950s and 1960s for just this purpose. By mounting several playback heads on a circular gear that could rotate at an arbitrary speed either against or with the tape to be played back, the temporal and frequency information could be independently controlled. For example, to pitch-shift the sound on a tape up an octave without affecting its playback time, one could set the recorded tape to pass over the rotating gear at its normal (recorded) speed, and the playback gear assembly would rotate against the direction of the tape such that the playback heads alternately passed over the tape with a relative velocity twice the normal. The composer Karlheinz Stockhausen, in fact, used this device a great deal.

Plug-ins that provide independent time- and frequency-scale modifications are usually based on an algorithm called the *phase vocoder*, which itself is an improvement of Homer Dudley's original Channel Vocoder, invented at AT&T Bell Laboratories in 1940.

The phase vocoder is called an *analysis/synthesis* routine, because it works by splitting the time-domain signal both vertically in frequency and horizontally in time. The sound file is split vertically into a series of frequency channels (like low-bass, mid-bass, midrange, mid-treble, and so on), it is *windowed* into adjacent time-domain chunks, and a Fast Fourier Transform (FFT) is performed on each frequency channel for each window. Performing the FFT takes us into the frequency domain; it yields a bunch of numbers that reflect the magnitude and phase of the frequencies that are present in the signal. We can then play with these numbers before performing the reverse transform, the Inverse Fast Fourier Transform (IFFT), and reassembling the frequency channels and time windows however we wish, to take us back to the time-domain and produce a time-stretched or frequency-shifted sound.

Such plug-ins can vary greatly in their specific implementations, but two controls you might encounter are the FFT size and the window length. We won't get into all the technical details here, but both settings can quite audibly affect the sounding result, and both controls involve trade-offs. The larger the FFT size is set, the more precise the frequency resolution, but the lower the corresponding temporal resolution can be.

Pitch Correction

Pitch-correction algorithms are relative newcomers to the plug-in scene, but they can be a very powerful and time-saving tool if not overused. They attempt to alter the perceived pitch of individual notes on a recording, and they are most often applied to vocal tracks. The user often begins by specifying a target tuning system grid, and the plug-in then quantizes notes that lie outside of this grid to the nearest "correct" pitch class.

Most people either love them or hate them, because, like any FFT-based analysis/resynthesis algorithm, their overuse (or misuse) can impart a characteristic "warble" or "gurgle" to the sound. If you are going to use a pitch-correction plug-in, practice and experiment while listening carefully to minimize these artifacts as much as possible. And keep in mind whether using a pitch-correction plug-in might adversely remove affective, intentional tuning variations from a recording before deciding for sure to use one.

Perhaps the most well known pitch-correction plug-in is AutoTune from Antares (http://www.antarestech.com; see Figure 5.15).

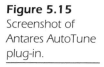

Figure 5.15
Screenshot of Antares AutoTune plug-in.

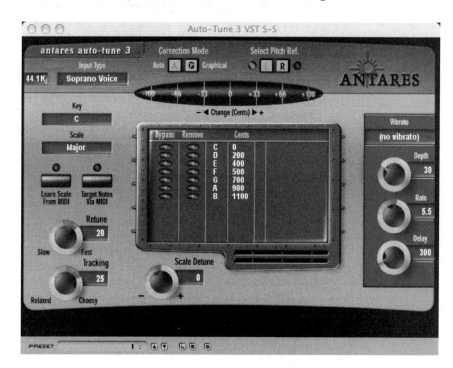

Pitch-correction algorithms, like noise-reduction and sound restoration algorithms (as we will see in the next section,) are generally used more for reparative processing than as a purely musical effect. As such, they can lend themselves, often unfortunately, to being used in "cruise-control" mode. Consider the product description from one such freeware pitch-correction plug-in, which promises to "[save] hours...on recording just to improve minor imperfections" and "satisfy your wishes in no time."

I do not mean to say that pitch correction cannot be a great time-saver when recording and mixing. I would submit, however, that given the choice between repairing a bad take and saving time versus recording until it is right, you should choose the latter. With that said, sometimes the luxury of time is not available, and pitch-correction plug-ins can effectively smooth over minor tuning problems in a recording if used sparingly.

Noise Reduction and Audio Restoration

Like pitch-correction plug-ins, both noise-reduction and audio-restoration plug-ins are usually used to "repair" sound rather than as a purely musical device. Indeed, these algorithms bring enormous sound-processing power to the DAW: Noise-reduction plug-ins can literally remove most unwanted noise from a recording, and audio-restoration plug-ins can remove all kinds of artifacts—from vinyl LP pops and scratches to digital fuzz and clicks—from sound files.

Most noise-reduction plug-ins work in the frequency domain, so they can take some practice to get to work properly and without leaving too many artifacts. Some restoration algorithms work in the frequency domain as well, but many fixes, like removing single-sample pops from a sound file, are best performed in the time domain.

Figure 5.16a illustrates a popular noise-reduction plug-in from Digidesign called Digidesign Intelligent Noise Reduction (DINR), and Figure 5.16b is a screenshot of BIAS SoundSoap Pro restoration plug-in.

Amplifier and Microphone Modeling

Modeling plug-ins attempt to impart the sonic characteristics of well-known amplifiers, loudspeaker cabinets, and microphones to an audio track. For example, upon specifying the type of microphone used to make

Plug-Ins

Figure 5.16
(a) Screenshot of Digidesign's DINR noise-reduction plug-in; © 2004 Digidesign, a division of Avid Technology, Inc. All rights reserved. Reprinted with permission of Digidesign; (b) screenshot of Bias' SoundsSoap Pro audio-restoration plug-in.

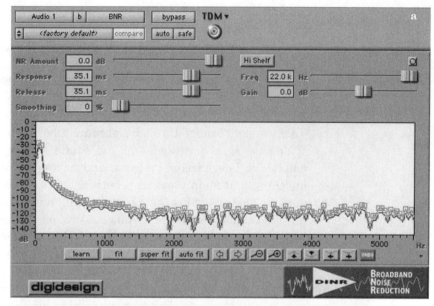

a particular recording, a microphone modeler can apply the measured transfer function of another specified microphone to the track, making the audio sound more like it was recorded with the target microphone. Or an electric guitar track that was recorded from an inexpensive speaker cabinet can be made to sound like it came from a larger, higher-quality cabinet. Amplifier modeling plug-ins can also, to some extent, make sounds recorded through a solid-state preamplifier sound more like they were instead recorded through a vacuum tube preamplifier.

You might ask why should you spend money on more expensive amplifiers, loudspeaker cabinets, and microphones, when you can use anything and simply model a better, more expensive device using a plug-in. These plug-ins can be very useful, but the effect of their modeling of a target microphone's transient response, for example, can be limited. For example, if the diaphragm of a low-quality microphone cannot move fast enough to record the sharp attack of a snare drum or the high-frequency components of a violin, then applying a transfer function to "bring out" what was not recorded in the first place will be of little value. With that said, however, these plug-ins can be quite effective at the tonal coloration of audio tracks, particularly when used on the voice.

Multieffects Processing (Chaining Effects Together)

Multitrack sequencers generally allow several plug-ins to be applied to each track. Remember when we spoke earlier about signal flow in a channel strip? It can become particularly important here: Clearly, the order in which we insert plug-ins to a track can affect the sounding result. Compare in your mind, for example, the difference that order makes when inserting an echo plug-in followed by a reverb plug-in. If we reverberate the echoed signal, then all echoes are reverberated and are confounded within the reverberant space. That is, the echoes themselves can interact within the reverb algorithm when creating indirect sound. If instead we echo the reverberated signal, the individual echoes cannot possibly interact within the reverb algorithm, because the reverb is not even "aware" of the existence of the echoes.

Several effects-chaining meta-plug-ins are available for use when using a large number of plug-in effects on a track. These are plug-ins that themselves act as managers for other plug-ins on that track, allowing users to graphically specify the precise routing matrix and signal flow for the plug-in effects on a track.

Miscellaneous and Experimental Effects

One of the exciting avenues enabled by plug-ins is the limitless number of audio effects that are possible. Virtually any processing algorithm that can be imagined can be implemented in software as an audio plug-in. The world of experimental sound processing, previously only accessible using more complex software synthesis programming languages and environments, is now easily accessible on the DAW using plug-ins.

A particular plug-in worth mentioning that is somewhat difficult to categorize is the BBE Sonic Maximizer. Available for years as a family of analog, rack-mounted processors, the plug-in version has been written to sound as much like its analog predecessors as possible. According to BBE, "[O]wning the Sonic Maximizer Plug-in is the equivalent of having a rack full of BBE 882i units with automation," which once again bespeaks the value offered by the modern DAW.

The BBE Sonic Maximizer attempts to restore the original phase relationships present in a recording that are often lost when amplified and reproduced over a loudspeaker. Often applied in the final or mastering stage of a mix to make it sound more focused and clear, correcting this phase distortion typically yields crisper transients.

One can envision almost anything as an audio plug-in: a vocal gender changer, a software synthesis environment, or even a scripting language for automated mixing. In a way, we have only begun to scratch the surface of what is possible.

Virtual Instruments

Virtual instruments are another family of plug-ins. As mentioned, virtual instruments emulate acoustic or electronic instruments through sampling or software sound synthesis. They come in two flavors: standalone programs that do not run in a host environment, and plug-ins, which exclusively work inside a DAW host program like a two-track editor or multitrack editor-sequencer. Each of these has its own advantages; many users, however, find the convenience of running a virtual instrument inside a multitrack editor-sequencer particularly appealing, because it is not necessary to import the resulting sound file into the mixing environment as a separate step.

When a virtual instrument plug-in is enabled, the graphical user interface provides control over the instrument's parameters. The instrument is typically played with a MIDI controller, like a piano-style keyboard for example, and the audio output of the virtual instrument is instantly placed onto the corresponding audio track.

Several kinds of virtual instrument plug-ins are available. These include software synthesizers, drum machines, and samplers. Some software synthesizer plug-ins even incorporate two or more of these functions within the same plug-in.

Synthesizers

Virtual instrument synthesizers usually attempt to mimic the functionality and appearance of traditional hardware synthesizers, from vintage analog models to newer digital keyboards, although their designs can just as well be constructed completely from scratch. Modular analog synthesizer plug-ins in particular can eliminate some of the difficulties of working with the hardware counterparts, like frequent de-tuning and tangles of patch cords. And the multitimbral and multiphonic limitations of hardware synthesizers can be overcome by relying on the fast processor speeds of modern DAWs.

Drum Machines

Like software synthesizers, drum machine plug-ins can either emulate the sound and appearance of famous hardware models, or they can be of a completely new design. Drum machines are synthesizers that are optimized for producing percussion sounds, and they incorporate pattern arrangers and sequencers to aid in the composition of percussion loops and grooves.

Samplers

Samplers are traditionally either incorporated into a keyboard synthesizer or a controllerless rack-mounted unit. They can load sound recordings into memory for modification, editing, and instant recall and playback. A sampler plug-in does the same thing but costs far less because it relies on the DAW's host computer for all of its operations. Furthermore, because it is a plug-in, its audio output can be immediately captured and placed into an audio track within a mix.

Musical Audio Effects Processing

Given the foregoing technical dichotomy of audio plug-ins, we are now in a better position to assess a more general, high-level aesthetic categorization. It may be easy to get bogged down in the sea of available plug-ins, but that is less likely to happen when you start with the musical vision of the sound inside your head and attempt to realize it concretely. Just as traditional pen-and-paper composers can write what their mind's ear can hear, so can any musician mix and produce audio from an internal aesthetic impetus. And just as traditional composers sometimes arrive at the printed notes on the page through a process of trial and error, often the result of many delightful musical surprises and happenings along the way, so too can trying different audio effects often spur imagination and interact with the other tracks in a mix in new and unimagined ways.

I like to think that audio effects plug-ins naturally fall into one of three broad aesthetic categories: effect/affect, blending, and emulation/illusion. Whenever processing audio (whether with a plug-in or any kind of sound editor), I find it helpful to always remain cognizant of which of these things I am trying to accomplish.

Effect/Affect

The simple processing of an audio signal can do a lot of things: It can create a barely noticeable difference in the sound, or it can yield a result that is entirely dissimilar to the original audio. Sometimes the simplest *effect* can evoke a particular *affect*. And while the processing of audio can assume any of a myriad processes, from pitch-shifting to flanging, it is at its heart a musical process that quite simply lets us sculpt a sound, sometimes even in isolation from its context, as our intuition directs us.

At its heart, I think of this notion of effect/affect as musical processing for its own sake. When applying plug-in effects to a sound in this mode of thinking, I am reminded of ornamentation in notated music. Is a trill on a particular note in that piano sonata entirely necessary? Usually not; the music would still stand on its own of course. Its an effect that can help evoke an affect.

Consider the idea of *transparency* when using plug-ins in this way. Is the effect supposed to stand out in the mix as an audio effect, or should its use be transparent to the listener?

Blending

Another major way in which audio effects can be used is to blend two or more elements together. For example, using effective equalization, we can make two tracks work better together and sound more distinct. Using various sonic morphing techniques, we can smoothly transition from one sound to another.

One particular category of blending frequently encountered in electronic and electroacoustic music is the attempt to meld natural, organic sounds with synthetic sources. Consider, for example, using a filter to color the sound of wind and then harmonizing it with a software synthesizer, or perhaps blending synthetic and natural sounds by placing them in the same acoustic space with a reverb plug-in. Or we could synthesize subharmonics to "beef up" a recording of a bass guitar to help it blend with the rest of the mix.

Another area in which blending is encountered is music for a prerecorded mix with one or more live instruments, like a piece for electric guitar, live processing on a computer, and computer playback of a precomposed accompaniment. The composed and controlled interaction among the live performer and the computer can, using carefully sculpted audio effects, create a broad spectrum of musical blending of the elements, from entirely dissimilar to the embodiment of one meta-instrument. As the old advertisement touted, "Is it live, or is it Memorex?" A skillful approach to sonic blending can lead the listener to question which sounds are being created in real time and which have been precomposed and mixed.

Again, no one plug-in or digital audio effect is the key to successful blending of musical elements. Trial and error, patience, practice, and intuition can lead the way in finding the right effect or effects to yield the best contextual blend.

Emulation/Illusion

The emulation of acoustic spaces, motion, physical acoustic phenomena, and organicism forms a third broad aesthetic category of digital audio effects. Using a reverberation processor, composers can virtually place individual sounds or an entire mix into another acoustic environment, as we previously mentioned. In this case, the composer is attempting to emulate the production of sound in another space, to, in essence, create a sonic illusion, using digital audio processing.

We can also emulate the movement of sound using audio effects. For example, careful application of surround panning in a 5.1-channel mix can create the illusion that sound is moving in space. This can be a very powerful and engaging musical device if done well and not overused.

The power of the modern DAW can also be used to emulate acoustical phenomena like Doppler shift, which can in turn help create virtual sonic illusion. If you have ever perceived the pitch of an ambulance to rise while the vehicle approached you and subsequently decrease while it drove away from you, then you have of course experienced Doppler shift. With a DAW, sound designers and composers can impose the same effect onto a recording of any sound to emulate the actual acoustical phenomenon.

Consider also the emulation of mechanicism, of "machineness." Applying a plug-in that synthesizes vinyl hiss and pops onto a digital audio recording reminds listeners of the recording medium, even if it is only emulated. Or consider using a vocoder plug-in to make a human voice sound like a robot, or a bandpass filter that makes a voice sound as if it were recorded through a telephone. An imparted illusion that reminds us of the machine can be musically quite profound in some contexts. And an imparted illusion that recalls a specific machine, like a plug-in that models the sound of a particular vintage microphone, or a particular guitar amplifier cabinet, can be both technically useful and sonically evocative, perhaps even engaging the composer's memory of another musical experience.

Emulation and illusion can also be elicited through the genesis of new material. In the context of plug-ins on the DAW, this can take the form of software synthesizer and sampler plug-ins. This may seem quite obvious, but I should state it here for completeness. Sound synthesis can be used to mimic acoustic instruments and phenomena, or it can of course be used to create artificial or physically impossible instruments. We talk about this more in a later chapter, but consider examples of the illusions of reality that can be created using samplers and software synthesizers: We can synthesize a realistic-sounding string orchestra, the smash of breaking glass, or the hits on a bass drum. Even though none of these acoustic events actually occurred, the illusion of their occurrence was created synthetically.

A final broad musical intention of emulation and illusion lies in the humanizing, or "organicizing," of sounds, which is of course the opposite of attempts to emulate machines. After the advent of the drum machine, many such machines and MIDI sequencers began including a "humanize" function to make the computer-perfect timing of beats a little more imperfect, and hence to impart to them a form of "humanness." Digital audio effects can now be applied to synthetic sounds to do just that as well. By slightly detuning the sound of a synthesized voice or imposing slight timing changes to an audio recording of a synthesizer sequence, for example, a sense of organicism—the illusion of created imperfection, but in more sophisticated ways—can often be evoked.

Summary

It is important to be aware of why you want to use a particular plug-in or audio effect. Consider the implications to the surrounding musical context, thinking both linearly in time and amplitude and vertically in frequency. And finally, if you cannot find a plug-in to do what you want, why not write your own, or work with someone to help you realize your idea?

Plug-In Formats

I hope you will pardon the diversion from aesthetic matters, but without the aesthetic/musical goal, none of the technical matters. It is fundamentally important to be constantly cognizant of both the technical and the aesthetic when using the DAW, or working in any kind of studio for that matter, so that technical hurdles do not impede the musical goal.

We began our discussion of plug-ins by taking a tour of them in terms of their technical and specific musical functionality, and then we grouped them into three broad aesthetic categories. We can also group them according to the specific protocol by which they work.

Just as sound files can be created in any of many different file formats, so too can plug-ins be created in various formats. However, unlike sound file format conversion, transferring a plug-in from one format to another is not as simple as saving it to a different format!

The specific format a plug-in assumes determines much about its functionality and usability. Some formats, like DirectX, only work on PCs. Other formats, like MAS and Audio Units, only work on Mac OS X. Most formats allow real-time operation, meaning that you can hear the effect of the plug-in on a track as soon as it is invoked. That is, the delay between input to and output from the plug-in's transfer function (remember those?) is more or less imperceptible, or at least psychoacoustically acceptable. Some formats, like AudioSuite, on the other hand, do not allow real-time operation.

We first briefly discuss each of the currently available plug-in formats, and then we examine a case study of creating a simple DAW plug-in in one popular format as an example. Keep in mind that plug-ins can often function in both two-track sound file editors as well as the most elaborate multitrack mixing environment. All that is usually required is to install copies of the desired plug-in into the application folders of each program, although some formats allow all plug-ins on a DAW to reside in a common folder.

The signal-processing computations required by plug-ins can be performed either on the DAW's host computer (*host-based plug-ins*), or they can be sent to dedicated hardware accelerators (*accelerated plug-ins*). The advantage of host-based plug-ins is that they do not require the purchase of external processing hardware, which can be as or more expensive than the host computer; however, using accelerated plug-ins means that a DAW can simultaneously run a larger number of plug-ins, but with the expense of additional hardware.

VST

In 1982, Charlie Steinberg started a company in Germany that began making a MIDI sequencer for the Commodore 64 computer, and then the more powerful Atari line of computers. By 1989, Steinberg's self-named company released a program that we have already talked about: Cubase. In 1994, he created one of the first audio plug-in formats, which he dubbed Virtual Studio Technology, or VST. The VST standard was officially unveiled at the 1996 Frankfurt Musik Messe. Today, hundreds of cross-platform VST plug-ins are available from both Steinberg and third-party developers—many of them completely free—and other programs besides Cubase support the VST plug-in framework. One of the reasons why so many VST plug-ins are available today, and in particular so many free ones, is that the specification is an open standard, meaning that anyone can download the source code for the specification and begin creating VST plug-ins using the supplied templates. In fact, a ton of free, user-submitted VST plug-in source code is available on the Internet.

The original VST specification has since branched out, leading to the development of the VST Instruments framework and now VST System Link. The VST Instruments specification allows "virtual instruments," like software samplers and synthesizers, to be developed within the framework of a plug-in. VST System Link, although a new technology, shows powerful potential for the modern DAW. Using this framework, the many processor-intensive signal-processing tasks required by multitrack mixers and plug-ins can be distributed among two or more computers. Even Macintosh and PC computers can be integrated into one conglomerate meta-DAW, with their shared processor power increasing the effective performance of the total system.

TDM

Digidesign, founded in 1983 and the makers of Pro Tools software/software systems for both Macintosh and PC computers, has created four proprietary plug-in formats: TDM, HTDM, AudioSuite, and Real-Time Audio Suite (RTAS). Digidesign's Time-Division Multiplexing (TDM) technology allows plug-ins written in this format to be run on dedicated Pro Tools TDM signal-processing hardware. That is, the DSP computations required to realize the plug-in's algorithm are "farmed out" to one or more host TDM cards.

AudioSuite

AudioSuite plug-ins are non-realtime effects that can be applied to a selection of audio or to an entire file. Because they operate outside of real time (meaning you must wait while the effect is applied to the file), they conserve the DSP horsepower of the host DAW computer for other mixing and plug-in tasks. However, also because they are non–real-time, they are best-suited for one-time operations that need not run in real time, like normalization, gain change, noise reduction, and DC-offset removal.

Real-Time Audio Suite (RTAS)

Digidesign's Real-Time Audio Suite (RTAS) is a format that, like the VST specification, allows the DAW host computer to perform the signal processing required by the plug-in. A variety of RTAS plug-ins is available, but because they are dependent on the host, the number of available simultaneous plug-ins is directly related to the processor speed, memory, and disk speed of the computer. On the other hand, for this very same reason, they can form the basis of a more cost-effective DAW-based studio that is not dependent on external processing hardware, as is the case with TDM plug-ins.

HTDM

Hybrid Time-Division Multiplexing (HDTM) plug-ins are host-based plug-ins that run on Pro Tools TDM systems. They send all processing computations to the DAW's host computer, freeing up additional processing horsepower for dedicated TDM signal-processing cards.

MAS

Mark of the Unicorn's MIDI-sequencing program Performer, released in 1985, was one of the first music programs written for the then new Macintosh computer. Several years later, with the development of their flagship DAW multitrack software Digital Performer, they produced a plug-in specification known as the MOTU Audio System (MAS) format. The MAS format is unique to and supported only by Digital Performer software at present.

DirectX

DirectX is a general-purpose multimedia plug-in format from Microsoft for use only in Windows operating systems. Introduced in

1995, the format has grown to include a large number of audio effects plug-ins, many of which are free. Unlike other plug-in formats, DirectX is a large collection of application programming interfaces (APIs) that act as a mediating layer between application software (like multitrack mixers) and the specific hardware connected to a computer (like a DAW audio interface), minimizing or eliminating the need for plug-in developers to write hardware-specific drivers and interface code. One of the technologies that DirectX supports is an audio plug-in format for Windows audio software, and much information is available online to assist DirectX developers.

LADSPA

The Linux Audio Developer's Simple Plug-In API (LADSPA; http://www.ladspa.org) is available for DAWs based on the Linux operating system. Remaining true to the open-source software movement on which Linux is based, LADSPA plug-ins are free and completely open, which can be a great benefit to new developers in learning how the format works. A large number of Linux host applications support the LADSPA format.

Audio Units

Apple Computer's Audio Units (AU) audio plug-in format is completely integrated into the MacOS X operating system as part of its Core Audio technology. The most recently developed format, AU enjoys the hindsight benefit of Apple's examination of all other plug-in formats. AU provides excellent graphical user interface design capabilities and features, and it is documented quite well.

Premiere

Adobe's Premiere video-editing software features a proprietary plug-in format for both video and audio processing. Support for Premiere audio plug-ins has waned, and the format is not supported by most major professional-level DAW software anymore.

Plug-In Wrappers

Plug-in wrappers are quite powerful: They allow a plug-in written in one format to be used in a program that only supports another format. For instance, the AudioEase VST Wrapper for MAS allows the vast number of VST plug-ins to be used within MOTU's Digital Performer software, which does not natively support VST plug-ins.

Each plug-in format specification includes a unique manner of invocation (that is, how an instance of the plug-in is actually created by the host program when the user selects that plug-in) and functional calls in software code. A wrapper is a kind of "translator" that converts the manner of invocation and the particular calls that one plug-in format expects to see into those that another format expects.

Creating DAW Plug-Ins

In the early days of the analog recording studio, it was not uncommon for mix engineers and technicians to routinely maintain, develop, and create new gear—audio oscillators, hardware effects processors, amplifiers, and so on. At least a rudimentary understanding of the practical side of electrical engineering was required to maintain a studio.

To the extent that software has replaced hardware in the modern DAW-based studio, an analogous phenomenon is happening: Musicians and computer programmers of widely varying skill levels are creating DAW plug-ins. And because hardware is not involved, the development cycle can be considerably shorter and less expensive in most cases.

Granted, developing a plug-in may not be for everyone, and if you know you are not at all interested in doing so, you can of course skip ahead. And a fundamental prerequisite, at least for developing your own plug-ins from scratch, is a working knowledge of object-oriented programming and an understanding of at least the introductory audio signal processing that we have covered. But as I've said before, and as my teachers of computer music often told me, it is important to be a developer when using music technology—not necessarily a software developer, although that is one example, but a developer of *tools*. In many ways, the extent to which we assume the role of tool creators dictates the very same level with which we exceed being users whose needs may be dictated by the technology, and hence the system of thinking, that we use when creating our music. Not that everyone needs to develop plug-ins to be a DAW power-user, but it is one way to get your feet wet if you are so inclined.

The Easy Way

Perhaps the simplest way to create a DAW plug-in is graphically. Using Pluggo from Cycling '74, for example, users can create graphical depictions of audio processes in their Max/MSP software synthesis program and instantly turn them into VST plug-ins. Other "brew-your-own" plug-in creation tools are available for the nonprogrammer.

Plug-In APIs

The other way to create plug-ins is to code them using a programming language like C++. Indeed, this is the method used to produce most professional-quality plug-ins. The process begins by selecting the target plug-in format, registering with the sponsoring organization or company to become a developer (which may or may not be free), and then downloading the appropriate applications programming interface (API) for that format.

The API for a plug-in format includes source code and documentation for the specification and usually a few sample (and simple) plug-ins. Often the best way to start is to make a simple modification to the source code, compile it, and test out the resulting plug-in in a host program.

User Interface Issues

Plug-in formats typically provide a mechanism for automatically generating a simple graphical user interface containing sliders, knobs, and buttons, with little or no effort on the part of the plug-in developer. In fact, in most plug-in format specifications, you can create a simple plug-in without writing a corresponding graphical user interface; the program under which the plug-in is invoked, called the plug-in host, must then automatically generate one for you. However, most professionally developed plug-ins have more or less elaborate user interfaces that have been custom-written by their developers or by professional graphic designers. New code-generating user-interface design tools have recently become available to help simplify this process.

Case Study: Getting Started Creating a VST Plug-In

As I mentioned before, one way to get your hands dirty (actually, I think I said "feet wet," but you get the idea) with a DAW is to register to become a third-party developer, for example, in the VST format. If you're not comfortable with writing C++ code, however, it will be quite frustrating! You can pick up much of the required DSP along the way, but at least a basic working knowledge of C++ is required. And because VST plug-ins are written in C++ and can run on different platforms, it is remarkably easy in most cases to compile the same plug-in for both Macintosh and Windows operating systems.

After you register to be a developer, you can then immediately download the VST Source Development Kit (SDK), which includes source code for the VST specification (required to compile any VST plug-in) as well as documentation and a few simple examples. Upon downloading the SDK, you should be able to compile the code without modification on your computer.

Explaining the interconnections of the components of a VST plug-in is a bit beyond our scope here; the documentation provided to developers does that well. In fact, the fastest way to start writing a plug-in is to dive right in, reading the source code itself. I would prefer to just whet your appetite for the heart of a simple plug-in, the part where the signal processing of audio samples is actually performed.

Before defining a signal-processing algorithm for a plug-in, we need to create the object's constructor. Let's look at the constructor of a simple gain-changing plug-in called AGain that is included with the SDK:

```
AGain::AGain (audioMasterCallback audioMaster)
        : AudioEffectX (audioMaster, 1, 1)     // 1 program, 1 parameter only
    {
        fGain = 1.;                            // default to 0 dB
        setNumInputs (2);                      // stereo in
```

```
setNumOutputs (2);                      // stereo out
setUniqueID ('Gain');                   // identify
canMono ();                             // makes sense to feed both inputs
                                        //    with the same signal
canProcessReplacing ();                 // supports both accumulating and
                                        //    replacing output
strcpy (programName, "Default");        // default program name
}
```

In the constructor of class AGain, we can initialize the values of whatever variables were defined in the corresponding header. Notice the first few lines of code, beginning with fGain = 1.;, which simply initializes the variable fGain to a floating-point value of 1.0. All internal signal processing in a VST plug-in is performed with 32-bit floating-point arithmetic, so all internal parameters must be declared as floating-point numbers. Furthermore, all parameters can only assume a range of 0.0 to 1.0. (The user interface can freely scale these numbers to appear however you wish.) Notice also that we are defining two channels of input and two channels of output with the setNumInputs and setNumOutputs statements. The rest of the function calls are included with many or most VST plug-ins, and their functionality is explained in the documentation.

At the center of a VST plug-in is an audio signal-processing algorithm. The core of our gain-changing plug-in is a C++ method called processReplacing:

```
void AGain::processReplacing (float **inputs, float **outputs,
long sampleFrames)
    {
        float *in1  =  inputs[0];
        float *in2  =  inputs[1];
        float *out1 = outputs[0];
        float *out2 = outputs[1];

        while (--sampleFrames >= 0)
        {
            (*out1++) = (*in1++) * fGain;        // replacing
            (*out2++) = (*in2++) * fGain;
        }
    }
```

Note that this method is passed a pointer to a pointer to the input samples to the plug-in (that is, a location in computer memory where the samples contained on a track are stored), a pointer to a pointer to the output samples (where the plug-in will place the samples at the conclusion of the computation), and the number of *sample frames* on which to operate at a time. A sample frame, sometimes called a *block*, is a chunk of audio samples grouped together. The plug-in or host program can define how many samples to pass to the plug-in at a time for processing. (Some simple processes, like a gain change, can operate on very small block sizes, while other more involved algorithms necessitate working on larger blocks.)

The interesting part of the process (well, okay, it's not really that interesting) is the *while* loop. This loop just sends the attenuated input samples to the output buffer. The input samples are attenuated, because they are multiplied by the variable AGain, which we initialized to 1.0 in the constructor.

What if we want to change the value of fGain dynamically? That's where the plug-in's user interface comes in handy. This simplest of plug-ins does not include a user interface, so it is up to the VST host to define one. Another method of class AGain, called getParameter, sets the value of fGain dynamically whenever the user moves the corresponding slider in the generated interface shown. Just compile the complete project, and now we have a complete working plug-in.

Creating plug-ins in any format involves much more than the surface we have just here scratched. But I hope that, if you are so technically and musically inclined, this foregoing discussion might spark your interest in some way.

Summary

The world of plug-ins is exponentially growing, in terms of both quality and quantity. And many plug-ins blur the rather rudimentary categorization given to them here by either encapsulating multiple functions into one plug-in or by featuring experimental, previously unavailable audio effects processing. Indeed, the very conceptualization of the plug-in as an audio analyzer, effects processor, and synthesizer, coupled with the sheer open-ended functionality they bring to the DAW guarantee both their deserved place within the modern audio workstation toolbox and the very usefulness of the computer as a musical instrument.

For Further Study

Bernardini N, Rudi J. "Compositional Use of Digital Audio Effects." *Journal of New Music Research* 2002;31(2):87–91.

Collins ME. *Professional Guide to Audio Plug-ins and Virtual Instruments*. New York: Focal Press; 2003.

Moylan W. *The Art of Recording: Understanding and Crafting the Mix*. New York: Focal Press; 2002.

Patten J, Recht B, Ishii H. "Audiopad: A Tag-based Interface for Musical Performance." *Proceedings of Conference on New Interface for Musical Expression (NIME '02)*. Dublin, Ireland, May 24–26, 2002.

Prager M. "The Story of VST: Interview with Charlie Steinberg." http://www.steinberg.net/en/community/world_of_vst/vst_story.

Risset JC. "Examples of the Musical Use of Digital Audio Effects." *Journal of New Music Research* 2002;31(2):93–97.

Schroeder MR. "Natural Sounding Artificial Reverberation." *Journal of the Audio Engineering Society*, 1962;10(3):219–233.

Steiglitz K. *A Digital Signal Processing Primer*. New York: Addison-Wesley; 1996.

Wishart T. *Audible Design*. York, UK: Orpheus the Pantomime.

Exercises and Classroom Discussion

1. Illustrate graphically how a flanger plug-in works by drawing a sine wave in time on one graph and a time-varying delayed version of itself. Add the two together to show the cancellations and rarefactions, and explain what you think this will sound like. Try it out on your DAW.

2. Prove to yourself or the class the difference that operating order can make when inserting plug-ins to a track.

3. Register to be a developer for a particular audio plug-in format.

4. Make a list of as many plug-ins as you can find. Develop a scheme to categorize them, and place each plug-in into the category that best describes it.

5. What is an allpass digital filter? What are its primary features, and why is it used in many time-domain reverberation algorithms?

6. Compute the complete convolution of the ideal balloon pop with the impulse response given in this chapter. Does the result make acoustical sense?

7. Create a delay plug-in for your DAW with a single slider to control the delay time.
8. Create a multi-tap delay plug-in for your DAW with multiple sliders to control multiple delay times.

CHAPTER **6**

A Closer Look
at the
DAW Studio

Creating a low-end DAW can be as simple as buying an inexpensive computer, plugging headphones into the audio output jack, and downloading some free software. Or it can involve weeks or months of research into the perfect computer, operating system, audio interface, control surface, software, and loudspeakers for your intended use and potentially much more money.

DAW Hardware

At the core, however, most musicians, whether setting up a budget home studio or a more elaborate professional studio, will at least purchase a good computer, high-quality external audio interface, and loudspeakers to monitor the DAW's audio output in conjunction with some of the software we have discussed.

Computers

Because the host computer is the central hub of the DAW, its features and specifications can be crucial in the successful DAW-based studio. Two things a DAW really likes are a fast processor (or several fast processors in the case of multiple-processor computers) and lots of random access memory (RAM).

People often ask if Macintosh or PC computers are preferred for DAWs. There are two answers, and it is up to you to select the one that best applies to you:

- **It doesn't matter.** Great software and hardware is available for both platforms and for Windows, Macintosh, and Linux operating systems. You can make great music with any kind of computer.
- **It matters.** Clearly, the choice of computer and operating system matters if there is a particular program you want to run that is only available on a certain platform. It also matters if you are already more comfortable with a particular operating system or already own a computer.

One of the more exciting DAWs is the laptop-based DAW. Laptops can now be equipped with many of the same audio interfaces previously available only to desktop computers while taking advantage of the portability of the laptop. Equipping a fast laptop with a high-quality audio interface and headphones can create a complete portable multitrack hard-disk recorder and mixing studio.

Audio Interfaces

The second main component of the DAW is its one or more audio interfaces, which contain higher-quality A/D and D/A converters than are typically found inside an off-the-shelf computer. These enable the DAW to serve both as a high-quality hard-disk recording system and multitrack mixer.

Audio interfaces can differ greatly in terms of their features. They can also vary significantly in the analog and digital audio connectors they support for getting audio into and out of the interface. Various connectors are used to transmit digital and analog audio between audio interfaces and the rest of the world, and transmission *formats* are often (but not always) tied to a particular *connector*. As such, we should pause to consider the kinds of physical connections and formats available for transmitting audio signals, for those who are not familiar with them.

Common analog cables/connectors and formats. The most common connectors for transmitting and receiving analog audio signals found in audio interfaces are quarter-inch, eighth-inch, RCA, and XLR (see Figure 6.1). Quarter-inch connectors come in two flavors: tip-sleeve (TS) and tip-ring-sleeve (TRS). TS connectors, used to make instrument cables, can conduct unbalanced (two-conductor) analog audio signals. One part of the connector (the tip) is designated "hot" (+), and the other part (the sleeve) is tied to ground.

TRS connectors can conduct balanced (three-conductor) analog audio signals. The tip is designated "hot" (+), the ring is designated "cold" (−), and the sleeve is again tied to ground. Unbalanced cables offer greater immunity from environmental noise and signal interference by transmitting an audio signal both in phase and out of phase. The hot connector carries the in-phase audio signal, and the cold connector transmits the same signal but out of phase. Because most electromagnetic noise incident on the cable from nearby radios, televisions, and the like, creeps into both signal paths in phase, we can then subtract the cold from the hot down the road and effectively cancel out most of the noise. (See Figure 6.2.) Note that the resulting signal after subtracting cold from hot is twice as big as the original signal, but

Figure 6.1
Common
connectors: (a)
quarter-inch TS;
(b) quarter-inch
TRS; (c) eighth-
inch TRS; (d) RCA;
(e) XLR male and
female.

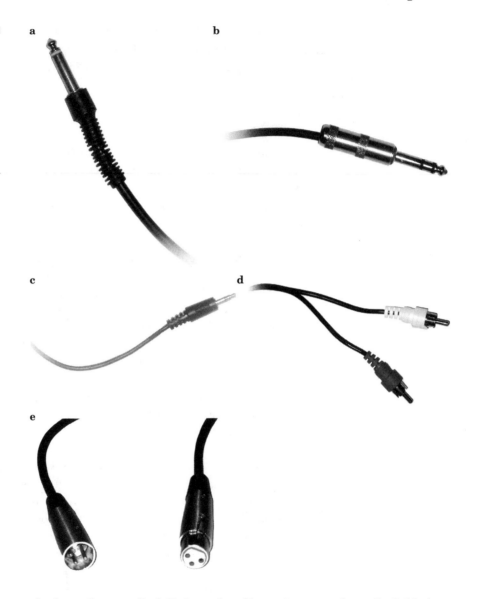

that's easily remedied. Balanced audio equipment takes all of this into account for us so that we don't need to think about it.

TRS connectors and cables can also be used for transmitting and receiving stereo unbalanced audio signals, but this is typically used only for front-panel headphone jacks on some DAW audio interfaces. Eighth-inch TS and TRS connectors are often encountered as well, most often as the built-in analog audio connectors on desktop and laptop computers.

Figure 6.2
Unbalanced
connections and
cables can be
used to help
prevent noise
from entering or
leaving the DAW.
(a) Original signal;
(b) signal in the
presence of noise;

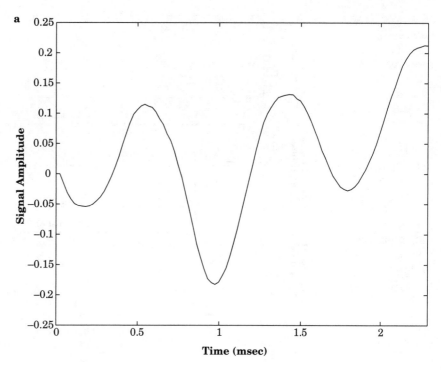

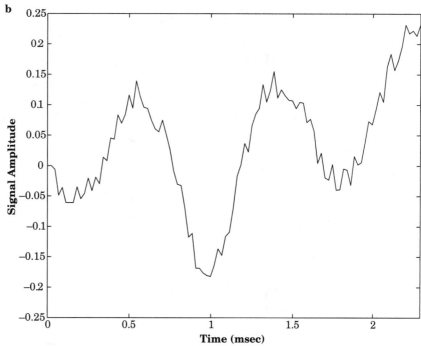

Figure 6.2
Unbalanced
connections and
cables can be
used to help
prevent noise
from entering or
leaving the DAW.
(c) balanced audio
system illustrating
in-phase ("hot"),
ground, and out-
of-phase ("cold")
signals; (d)
subtracting the
cold signal form
the hot signal
eliminates virtually
all incident noise,
since the noise is
in phase in both
hot and cold.

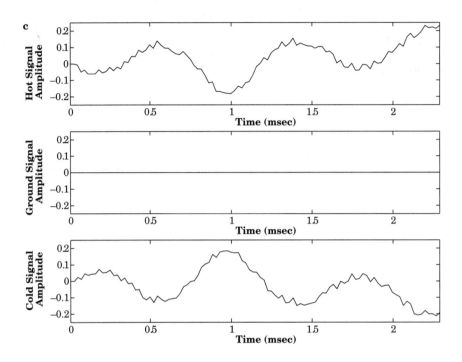

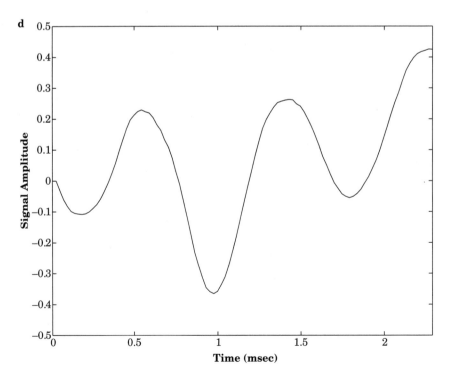

Analog audio can also be transmitted using RCA ("phono") cables, although they are limited to unbalanced audio signals. RCA connectors include a center "hot" (+) protrusion and a surrounding ground connector. They are most often encountered among budget-oriented DAW audio interfaces.

Cables made with 3-pin XLR connectors (XLR-3) can transmit analog balanced audio using three connector pins, labeled 1 (ground), 2 ("hot"), and 3 ("cold"). XLR cables offer the distinct advantage of being able to physically lock into place once plugged in: They are difficult to remove until a small release button is engaged.

Common digital cables/connectors and formats.   Most DAW interfaces can also exchange digital audio with other digital audio equipment. When working with mono or stereo digital audio, it is customary to use S/PDIF or AES/EBU cables and connectors, while ADAT, TDIF, and MADI are used for multichannel digital audio.

The Sony/Phillips Digital Interface (S/PDIF) standard allows digital audio to be transmitted over RCA cables. Because RCA connectors are relatively inexpensive, many budget-oriented DAW interfaces include one or more S/PDIF connectors. (S/PDIF data can also be transmitted optically, but that is less commonly encountered.)

The Audio Engineering Society/European Broadcast Union (AES/EBU) is a professional standard that allows stereo digital audio to be transmitted using standard 3-pin XLR connectors. AES/EBU connectors are usually found on higher-end audio interfaces.

The Alesis Digital Audio Tape (ADAT) protocol allows eight channels of digital audio to be transmitted simultaneously over a single optical cable terminated with TOSLINK ("Lightpipe") connectors. The Tascam (Teac) Digital Interchange Format (TDIF) similarly allows simultaneous transmission of eight channels of digital audio, but it uses DB-25 connectors. A major difference, however, is that the ADAT format can carry word clock (a synchronization signal), while TDIF does not. Thus, when using TDIF connections, a separate word clock connection must be made.

Another multichannel transmission format is the Multichannel Audio Digital Interface (MADI), which is capable of carrying up to 64 channels of 24-bit digital audio. Previously only found on high-end digital mixing consoles, some newer digital audio interfaces now include MADI connections. MADI can be transmitted using coaxial BNC or fiberoptic cable; it supports 64 channels of up to 48 kHz audio and 32 channels of up to 96 kHz audio.

Some audio interfaces employ proprietary formats using other connectors. For example, the MOTU 828 audio interface transmits digital audio using the IEEE 1394 "FireWire" protocol. Other interfaces can connect to expansion cards in laptop computers using CardBus connectors.

Choosing an audio interface.   Many audio interfaces for computers are available, and more are continually being developed. Many of the companies that make DAW software also make hardware audio interfaces designed specifically to work with their software, which can greatly ease installation, configuration, use, and maintenance.

Regarding the choice of an audio interface, I only offer two overriding principles: less is often more, and you usually get what you pay for. There is no need to pay for features you will never use; it's usually a good idea to purchase an interface and software that allows you to do what you want and maybe a little more, so there is room to expand. And regarding cost, some audio interfaces cost more because they have better-sounding, more expensive A/D and D/A converters or higher-quality connectors.

Audio system interconnectivity.   In the "old days," using an audio interface with a computer usually required installing a set of drivers specific to a particular software application in order for the software to communicate with the hardware. In fact, before programmers could sit down and write their application, they typically had to research the hardware-specific communications protocols employed by a particular audio interface. And so writing software that could work with a myriad of different interfaces often entailed writing a myriad of different drivers as well.

Thanks to technology like Steinberg's ASIO (Audio Stream In/Out), standard which has now been around quite a few years, communication between hardware and software can be mediated by a robust abstraction layer. This allows the hardware designers, who of course best understand the inner workings of their devices, to write ASIO-compliant communications drivers, for example, and ASIO then simply provides a common set of routines that can be involved by audio software designers. A similar interoperability goal has been embraced by the PortAudio project (http://www.portaudio.com), which is an open-source collection of routines that allows programmers to write software that can access the internal sound hardware on computers and read/write sound files. Software written using PortAudio can then be easily recompiled under various operating systems.

Other technologies, such as Steinberg's ReWire and Cycling 74's Soundflower (http://www.cycling74.com), allow audio programs to communicate directly with each other on the same computer. For example, a standalone software synthesizer might stream audio samples directly to a multitrack editor, or audio from an Internet radio station could be recorded into a two-track editor.

Loudspeakers and Headphones

Monitoring the audio output of a DAW requires a set of loudspeakers and/or headphones. Headphones offer the advantage of extended frequency response and potentially lower price (as well as the ability to mix at four in the morning without disturbing neighbors!), and they can be occasionally useful when mixing to provide another perspective, but mixes made entirely using headphones tend to sound overly expanded in terms of their dynamic range. Because the headphone's transducer lies in such physical proximity to the eardrums when headphones are worn, the effective sensitivity of our ears to variations in amplitude is much greater, which can lead to overcompensation when using dynamics processors and setting levels in a mix.

Monitor loudspeakers ("monitors") are available in passive and active varieties. Passive monitors require an external amplifier to operate, while active monitors feature one or more amplifiers packed inside the loudspeaker cabinet itself. Thus, passive loudspeakers accept only speaker-level (low-voltage, high-current) inputs, while active loudspeakers accept only line-level (low-voltage, low-current) inputs.

Mixes can be monitored using three basic configurations: nearfield monitoring, midfield monitoring, and farfield monitoring. Nearfield monitoring involves smaller loudspeakers placed in close proximity to the listener, while midfield and farfield monitoring involve larger loudspeakers placed farther away, respectively, from the listener. Each offers a different acoustical perspective on a mix, although many home-based project studios function just fine with only one set (or perhaps two different sets) of nearfield monitors.

Input Devices and Control Surfaces

Other hardware components that many DAW users find useful are control surfaces and various input devices. Control surfaces are simply devices that provide a more immediate, tactile interface to the DAW's functionality. For example, a mixing control surface can emulate the look and feel of a hardware mixer while affording the processing power and open-endedness of the DAW. This can provide ergonomic benefit as well as a sense of immediacy and familiarity, particularly to musicians accustomed to working in traditional studios.

Alternate input devices can similarly enhance the ergonomics, utility, and efficiency of the DAW. For example, a scroll-wheel input device can greatly increase the accuracy and speed of the fast-forward and rewind operations of a multitrack mixer by emulating a hardware jog shuttle.

Hardware Mixing Consoles

The addition of a hardware mixing console, whether analog or digital, can often be helpful if the DAW will be part of a larger studio or interfaced with other audio equipment like recording devices and microphones to help set input and output levels. However, because audio processing and mixing are performed completely inside the DAW, a relatively simple mixing console can typically suffice in a DAW-based studio, if you decide to use one.

Setting Up a DAW-Based Studio

Establishing a studio of any size from scratch can be a daunting (and potentially very expensive) task. As we mentioned earlier, though, DAW-based studios comparable in overall functionality to traditional console- and outboard-gear-based studios can be considerably less expensive. For the beginner especially, there is simply no better choice. For marginally more money than a so-called "studio-in-a-box" system, which typically includes a digital mixer with hard disk and recordable compact disc drive as the heart of a home studio, a high-quality project studio can be built around a DAW. And because the DAW is based around a general-purpose computer, the potential for expansion and upgrades can be much greater.

Four key steps are involved in setting up a DAW-based studio:

- **Determine the size and scope of the studio.** Is it a multiuser production facility or a single-user home project studio in your garage?

- **Select and acoustically treat an appropriate space.** A better-sounding space can lead to better-sounding mixes.
- **Purchase an appropriate host computer, audio interface, and software for the DAW.** Research your needs and choose what is best for you.
- **Determine necessary peripheral equipment.** Does the studio require a mixer control surface and lots of extra disk storage, or is the computer itself sufficient? What about microphones?

In each of these steps, it is crucial to maintain a budget and adhere to it as closely as possible. Although continually purchasing the latest equipment may be perpetually tempting, consider how you would incorporate the new equipment musically before buying. And as with purchasing a car, it is also a good idea to "test drive" any equipment before purchasing it.

Like any studio, the first step of determining the size and scope of a DAW-based studio dictates many choices one can make. For example, a low-end home studio might run fine on a slower, older computer, whereas a multiuser studio facility run for profit would require the latest equipment for clients. If a portable studio is desired, a laptop might work best, which means one much choose a laptop-compatible audio interface, or perhaps one specifically designed for a laptop.

The networking requirements of a studio must also be considered. Virtually all users will find Internet connectivity very useful, particularly for software updates and sending audio files. If you are a home user, you might consider getting a high-bandwidth connection, such as cable, DSL (digital subscriber line), DSL2, or satellite if they are available. Audio files can be very large, so a high-bandwidth connection to the Internet can be very useful for backing them up and transmitting them.

The second step, selecting an acoustically appropriate space, is also directly related to the size and scope of the studio. For the budget-conscious home user, a quiet bedroom may be suitable, or perhaps a two-car garage could be converted. Some great books are available that detail budget-conscious studio design techniques.

Next, research the most appropriate computer, software, and audio hardware system for your needs. The range of available options is continually growing to fit all budgets. Finally, unless you are a seasoned pro, consider limiting your initial purchases of peripheral equipment. Most users of DAW-based studios seem much happier with faster computers than excess outboard gear.

The DAW-Based Stereo Project Studio

Perhaps the most commonly encountered DAW studio type is the stereo project studio. We talk here about some simple things you can do to create a good mixing environment for such a studio, although most of these techniques apply to all kinds of project studios in general. When setting up any studio, one of the first items to consider is the location. Although "room-within-a-room" studios are very desirable acoustically, they are often not feasible economically. But aside from choosing the quietest room possible, many things can be done to help almost any room sound much better. The guidelines here, although by no means exhaustive, can at least help you get up and running with a budget DAW-based stereo project studio.

Why do we care about the acoustics of our mixing environment? Because it affects the way the resulting mix sounds! The ideal mixing location is a room that does not color the sound in any way, but such a room is architecturally impossible. The best we can hope for is to minimize the coloration effects of the room using various techniques. These techniques fall into four categories: *architectural modification*, *noise reduction*, *speaker placement*, and *acoustical treatments*.

The architectural modification of a room is usually the most expensive and time-consuming technique for minimizing coloration in a mixing environment, but there are ways to do it on a budget. Many people, for example, convert a spare bedroom or their garage into a DAW-based project studio. During the conversion, a simple but effective way to minimize the acoustical coloration of the room is to create non-parallel walls, which diminishes slap-back echo and standing waves. It is possible to create a great-sounding studio with parallel walls using acoustical treatments that we will talk about shortly, but many project studios benefit from the simple addition of extra drywall framed with 2 × 4 wood studs splayed at an angle of around 5° to 10°. Other modifications include creating a "floating" ceiling by dropping acoustical tiles below the existing ceiling (provided the ceiling is high enough) and splaying the studio's front wall forward slightly towards the ceiling.

Assuming you have chosen the quietest room possible for the project studio, it is imperative to keep as much ambient sound and noise out of the room as possible. External noise can be somewhat attenuated using acoustical treatments, about which we will speak in a moment, but the best thing to do is to prevent noise from being created within the room at all. And because the DAW is based on a computer that probably contains a fan to cool its internal components, DAW-based studios can be particularly susceptible to computer noise. Ideally, desktop computers should be placed in another room or closet, which of course necessitates extension cabling for the computer monitor, input devices, and so on. If this is not practical, a noise-control isolation box built specifically for DAW computers can be purchased (or you can make one yourself if you are so inclined).

Second, the physical locations within a room where monitors are placed can greatly impact the tonal characteristics of a mix. Any physical object nearby could potentially impede or reflect the transmission of sound waves emanating from the monitors. Also, the closer monitors are placed to a reflective surface, like a wall or especially a corner, the more bass-heavy their response becomes. They should not be separated too far apart, and they should be angled slightly inward toward the "sweet spot"—the location you normally sit while mixing.

Finally, acoustical treatments can be purchased or even homemade to greatly improve the acoustics of a DAW-based stereo project studio. Acoustical treatments fall into two basic categories: *absorbers* and *diffusers*. Absorbers are typically made of a foam-like material or acoustical fiberglass and minimize the ambient sound level in a room by absorbing acoustical energy. They are placed on walls and in corners to absorb accumulated sound waves and flatten the tonal characteristics of the room. One simple way to figure out where to place them on a wall is to clap your hands near your seated mixing position and listen for flutter/slap echoes and comb-filtering effects that may occur because the sound is bouncing between walls. With careful listening, you can discern which walls are the culprits and place absorbers on those walls. In professional high-end studios designed from the ground up, coloration effects like these are dealt with beforehand by modeling the acoustics of the room in software, but most budget project studio designers do not have (or generally require) this luxury.

A bass trap is a special kind of absorber that is placed in the corner of a room to minimize the collection of low-frequency energy that is susceptible to accumulate there. Low frequencies can be especially difficult to tame in a room owing to their long wavelengths. (Remember $\lambda = c/f$?) Bass traps in project studios are typically fastened to the corners of a room with an acoustical adhesive or supported by stands, and they can serve not only to tune the low-frequency coloration in the mixing room but also limit the transmission of bass frequencies to other adjacent rooms.

Absorber panels can also be made using 1x6 wood panels, dense fiberglass, and speaker fabric or burlap. While the fiberglass available at the local hardware store is not as effective as acoustical fiberglass, it does absorb sound pretty well, and the denser and thicker the fiberglass, the better the absorption. A budget absorber panel can be made quite simply by constructing a rectangular frame with 1x6 panels, stuffing dense fiberglass into it, and then covering the entire frame with speaker fabric or burlap. Be sure to apply a flame-retardant to the burlap covering, as burlap (and fiberglass) is highly flammable.

Acoustical diffusers, on the other hand, attempt to disperse acoustical energy in a room to minimize the buildup of standing waves and collections of acoustical energy. Standing waves can occur between any two parallel surfaces when a frequency is sounded whose wavelength corresponds to the distance between the surfaces. They can also occur when integer multiples of that fundamental frequency occur. Diffusers work by breaking up and redirecting the path of the sound waves. They can enlarge the mixing "sweet spot" by diffusing and delaying the early reflections that occur in a room; they can create a longer time between the direct sound and the early reflections, thereby enhancing intelligibility and presence while minimizing "mushiness" in a room.

Popular acoustical product makers include Auralex Acoustics (http://www.auralex.com) and Acoustical Solutions (http://www.acousticalsolutions.com). Some companies even offer "acoustics-in-a-box" solutions that feature an entire collection of acoustical products tailored to a specific studio. Much more can be said about creating a budget studio, and several excellent reference books are listed at the end of this chapter.

For Further Study

Davis G, Jones R. *Yamaha Sound Reinforcement Manual*. Hal Leonard; 1990.
Everest FA. *Sound Studio Construction on a Budget*. New York: McGraw-Hill; 1996.
McCartney T. *Recording Studio Technology, Maintenance, and Repairs*. New York: McGraw-Hill; 2003.
Shea M, Everest FA. *How to Build a Small Budget Recording Studio from Scratch*. New York: McGraw-Hill; 2002.
White P. *Basic Home Studio Design*. Sanctuary Press; 2000.

Exercises and Classroom Discussion

1. Clap your hands once in your studio as loudly as you can. Do you hear any reflections or coloration? Try popping a balloon, also.
2. Research various acoustical products for making your home studio sound better. Search the Internet for companies that sell acoustical treatments, and ask them for help in treating your studio.
3. Obtain the impulse response from a large acoustic space, such as a big church or auditorium, by recording the sound of a balloon popping while the space is as quiet as possible. Place a sound file virtually in that acoustical space by convolving the sound file with the impulse response of the space.
4. Construct an experiment to suggest the best acoustical locations in a large hall (that is, the spaces in the hall that exhibit the least coloration) by obtaining multiple impulse responses from various locations.
5. Why does placing monitors in corners of a room exaggerate their production of bass frequencies?

Surround Mixing and Monitoring

The first leaps from monophonic recording and playback to stereophonic audio reproduction must have been sensational to experience firsthand. Although credit for the first attempt to broadcast a two-channel audio signal is historically bestowed upon the self-taught aeronautical engineer Clement Ader (in 1881 at the Paris Exhibition), a theory of stereophonic recording and reproduction of phantom images was not fully developed until many years later. History had to wait until the 1930s for pioneers like Alan Blumlein, who patented several stereo recording devices and techniques, and engineers at Bell Laboratories, who developed a three-channel (Left-Center-Right, or LCR) sound-reproduction and amplification system for large auditoriums, to lay the early groundwork for the viability of stereophony as a commercially distributed and successful audio format.

And to the extent that recording and playback attempt to mimic or create sonic reality—whether through direct capture of acoustical space or artificial creation thereof—surround audio too provides another technological and aesthetic leap forward. Mixing audio in surround brings a wealth of creative possibilities to the modern DAW. Indeed, the direct placement of sound in three-dimensional space enables the creation of immersive musical environments and reinvigorates the musical notion of space—the sound stage—as a central mixing and compositional parameter.

Central to crafting a good surround mix on a digital audio workstation is creating as accurate a monitoring environment as possible—one that will accurately reflect the arrangement of loudspeakers during playback on the target system (whether a home theater, a car, a concert hall, or a movie theater) as closely as possible. And with the vast number of surround audio formats in use today, creating that environment can be somewhat confusing. We have already spoken about basic speaker placement techniques and acoustical room treatments for DAW project studios; these techniques, coupled with an understanding of common surround configurations and formats, can help you create a better surround mixing environment.

One of the best things about DAWs is the flexibility they offer in monitoring a mix, irrespective of speaker configuration. For example, with a relatively inexpensive eight-channel sound card and several loudspeaker monitors, one can monitor and mix in mono, stereo, LCR, quadraphonic, LCRS, 5.1, 6.1, 7.1, and 8.0. (Don't worry if that last sentence didn't make much sense; we'll get to those terms in this chapter.)

This chapter discusses the basics of how we hear sound in space, including a psychoacoustical overview of how the ear and brain work in tandem to help us localize sounds. Following a survey of aesthetic considerations prompted by surround audio and an examination of various surround configurations, we then turn our attention to various techniques for modifying, creating, and enhancing space in a mix—first using traditional techniques, and then by examining the concepts of ambisonics and binaural processing.

But once a surround mix has been made, what do we do with it? We conclude by discussing various means of presenting and disseminating digital music that are exclusive to surround audio.

Surrounded by Sound

Once sound pressure waves enter the ear canal, a wondrously complex series of acoustical, mechanical, and electrical processes and interactions begins in order to impart in us a sensation of hearing. The mechanisms by which we aurally "observe" sound enable us to perceive many things about the source to varying degrees of accuracy, including where it is coming from, how far away it is, whether it is moving, how loud it is, its timbral characteristics, its temporal characteristics, and of course something about its frequency content and pitch characteristics. All of these cues can be interpreted in our brains almost immediately to help us determine first if the sound presents a threat or warning by attempting to discern its source. (Is it a fire alarm? Is a tiger growling behind me?) As part of this innate perceptual mechanism, we are well equipped to discern not only the time and frequency characteristics of sounds, but also their spatial location and distance.

The psychoacoustics of spatial hearing is the subject of many ongoing research studies by neurologists and music cognitionists, and although much progress has been made, we still have a long way to go. We limit our discussion here to the classic "textbook" mechanisms by which we localize sounds that are well understood. These can easily be decomposed into three categories: *timing cues, level cues,* and *tonal cues.* Each of these cues is based on the physical fact that sound travels at a constant speed of propagation in air—around 343 meters per second, or 1,131 feet per second—as weall as how it interacts with the physiological construction of our heads and torsos. Just as two eyes are required for true three-dimensional vision, so are two ears required for three-dimensional hearing. These cues, working in tandem, help us to localize a sound source's *azimuth*[1], *elevation*, and *distance*, in order of roughly increasing difficulty.

Timing Cues

Unless the source of a sound is directly in front or in back of a listener, the acoustic pressure waves it generates reach one ear sooner than the other. The time difference between the arrival of the first wavefront of the wave in each ear, although extremely small, is enough to serve as a primary cue for azimuth perception.

This timing cue, discovered over a century ago by Lord Raleigh (1842–1919) as part of his Duplex Theory of sound, is customarily referred to as the interaural timing difference (ITD). Figure 7.1 illustrates the general idea. The ITD cue is most salient for lower frequencies—those below about 1.5 kHz—because the wavelengths of these frequencies are longer (remember $\lambda = v/f$ again?), and there is less ambiguity in the number of waveform periods have been delayed (that is, the phase difference between the ears is smaller.)

Figure 7.1
The Interaural Timing Difference (ITD) cue: each acoustical wavefront (for frequencies below about 1.5 kHz) from a sound source reaches the ears at different times.

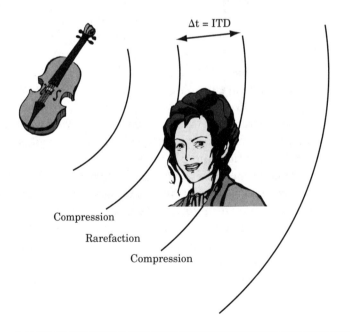

$\Delta t = \text{ITD}$

Compression

Rarefaction

Compression

1 The azimuth is the angle in the plane of the floor on which the listener is standing. In other words, the azimuth angle would indicate if the sound is coming from in front, behind, the left side, or the right side, or some angle in between, of the listener.

We are also pretty adept at localizing sounds in highly reverberant environments, thanks largely to the *precedence effect*. (In an "muddy" acoustic environment, we are innately able to temporarily block out the early reflections and reverberant wash of sound to help localize sound sources. According to the precedence effect, also called the Law of the First Wavefront or the *Haas effect*, we tend to localize sounds based primarily on the first acoustic wavefront to hit our ears; that is, our brains can group the direct sound and early reflections into one perceptual gestalt. The effect is prominent for early reflections that occur within the first 30 milliseconds or so of the direct sound, provided the early reflections are not more than about 10 dB louder. After about 30 milliseconds, the effect disappears, and the early reflections begin to be perceived as distinct entities, eventually turning into full-fledged echoes by around 50 milliseconds.

Another more subtle timing cue relating to the precedence effect involves the *Franssen effect*, caused by the interaural envelope difference (IED). The IED corresponds to the differences in the global amplitude envelope of a sound source as it arrives at the ears. According to the Franssen effect, if a short sine-wave burst is played in one loudspeaker and cross-faded into another loudspeaker placed some distance away, under certain conditions we tend to perceive the sound emanated from the first loudspeaker for the entire duration.

Level Cues

Level cues help the listener discern the azimuth and distance of a sound source. It makes intuitive sense that a sound source is probably located closer to us if we perceive it as louder. The difference in perceived level between the ears is called the interaural level difference (ILD, also called the interaural intensity difference, or IID), which serves also as a primary azimuth cue, complimentary to the ITD. Interaural level differences are primarily caused by diffraction of the incident sound waves off the listener's head; some acoustic energy bounces off the listener after reaching the first ear, and so less acoustic energy is available to reach the other ear, creating the ILD.

A large ILD can also be created when the listener's head casts a kind of acoustic shadow, blocking the other ear from receiving much acoustical energy at all (Figure 7.2). For high frequencies, particularly above about 1.5 kHz—just where the ITD leaves off, in fact—the ILD takes over. This is because higher frequencies, which have short-

Figure 7.2
The Interaural
Level Difference
(ILD) cue can be
created from
acoustic
shadowing.

er wavelengths relative to the size of head, are acoustically blocked from passing around the head.

Level cues also serve as a primary component in our perception of distance. The louder a sound is to our head, the closer it is more likely to be. Other level effects help as well, in particular the ratio of direct to reverberant sound. When a sound source is close by, we hear a much greater ratio of direct sound to reverberant sound; conversely, in a highly reverberant environment like a large cathedral, a sound can be far enough away that we hear virtually no direct sound.

Tonal Cues

Tonal cues, that is, the perceived frequency coloration of sounds, can also aid in localizing sound sources. One simple way is that the farther away a sound source is, the more low-pass-filtered it tends to sound. (Dynamic low-pass filtering is thus one classic technique of creating near-far space in a mix, in addition to varying the ratio of direct to reverberant sound, as we just mentioned.)

The outer folds—or pinnae—of the ears are shaped funny for a reason: They act as a space-dependent filtering mechanisms for incoming sounds. As a sound enters the outer ear, it bounces around a bit before entering the ear canal; in so doing, the tonal characteristics of the sound are changed. Some frequencies are amplified, while others are attenuated. This coloration serves as our primary elevation cue, but it can assist azimuth perception as well.

The way that each person's ears color incoming sound is slightly different. Treating the human head like a kind of "black box" acoustic processor, we can measure the effect of this coloration by placing a tiny microphone inside the ear and recording the tonal variations of incoming sound played from different locations in an anechoic chamber. Doing so will lead to a complete data set known as the subject's *head-related transfer function* (HRTF). See Fiugre 7.3. Applying an HRTF-based psychoacoustic processing program or DAW plug-in leverages this measured data to spatialize sounds virtually in three-dimensional space. This technique can be highly effective when a mix is played back over headphones.

Figure 7.3

A sample HRTF data set measures the tonal coloration that a particular listener's ears create for various sound source locations. Shown here are data for one listener's left-ear coloration for sounds at 0° elevation and (a) 0° azimuth directly in front; in the plane of the ears;

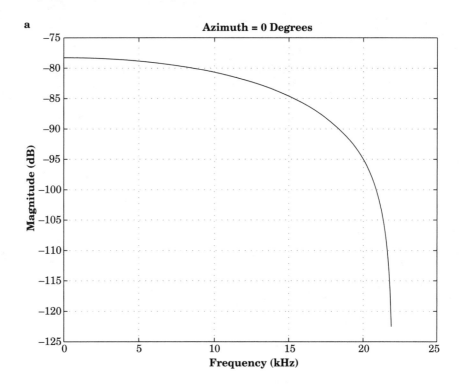

Figure 7.3
A sample HRTF data set measures the tonal coloration that a particular listener's ears create for various sound source locations. Shown here are data for one listener's left-ear coloration for sounds at 0° elevation and (b) 15° azimuth; (c) 30° azimuth;

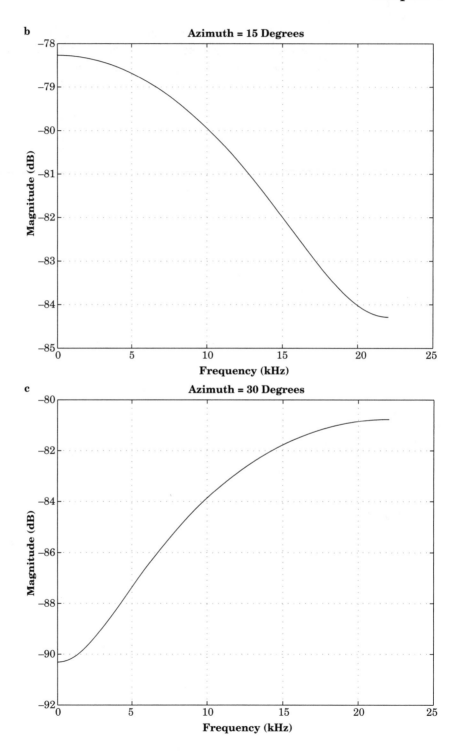

Figure 7.3
A sample HRTF data set measures the tonal coloration that a particular listener's ears create for various sound source locations. Shown here are data for one listener's left-ear coloration for sounds at 0° elevation and (d) 45° azimuth (due right); (e) 60° azimuth;

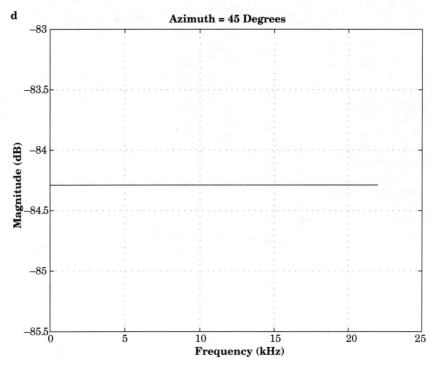

d

Azimuth = 45 Degrees

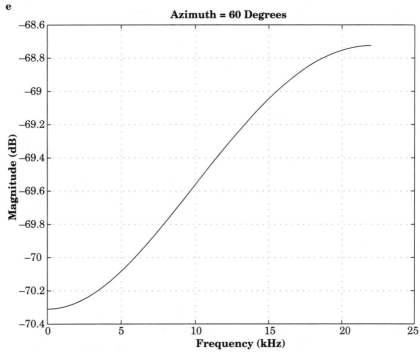

e

Azimuth = 60 Degrees

Figure 7.3
A sample HRTF data set measures the tonal coloration that a particular listener's ears create for various sound source locations. Shown here are data for one listener's left-ear coloration for sounds at 0° elevation and (f) 75° azimuth; and (g) 90° azimuth (directly behind). Data are taken from the compact Gardner-Martin HRTF set.

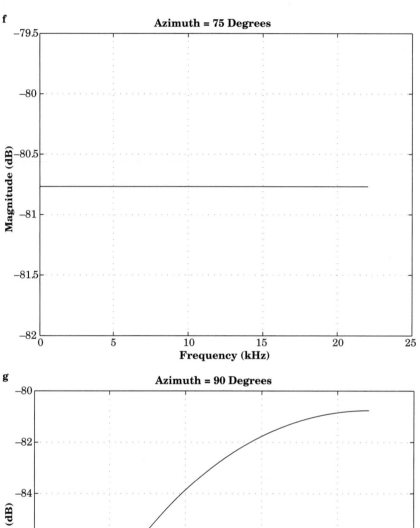

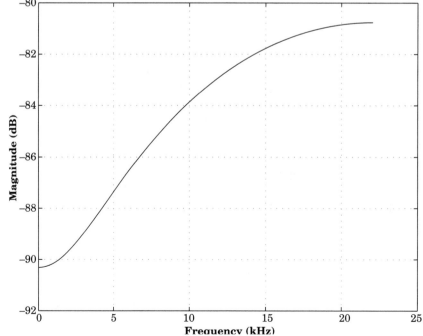

Of course, many other cues can help us perceive the location a sound source, including visual cues, head movement, expectation, memory, and so on. But with these basics out of the way, we can now get back to surround audio on the DAW, a technically interesting and yet simultaneously creatively rich area of study.

Surround Audio on the DAW

Creating music for surround playback involves many technical and aesthetic challenges and questions. Where should the extra loudspeakers be placed? How loud should a subwoofer be set relative to the full-range speakers when mixing? Where should this audio track be placed in the surround sound stage?

These are some of the questions we attempt to answer in this section as we examine surround audio on the digital audio workstation. To this end, we discuss rudimentary techniques for setting up a project surround studio and then briefly study each of the major historical, currently popular, and proposed surround audio channel arrangements and formats. But to navigate the complex maze of possibilities for surround mixing, we start by addressing some of the aesthetic considerations involving surround mixing and composition. That is, what are the creative reasons for your wanting to mix in surround? The brief space allotted here will, I hope, serve as but a starting point for those interested in further study of this subject. It is an area rich in musical history, and the sky is the limit for the future.

Aesthetic Considerations

Mixing audio, just like composing music, is an exercise in taming an infinite number of aesthetic choices. And mix engineers, just like composers, try to maintain a continual dialogue with the creative impetus behind a mix. It is indeed a process of almost constant self-questioning for many. Does this or that sound better? What, if anything, did inserting that plug-in here contribute to the mix? Is this piece too brief? Is that vocal line as intelligible as it should be? Is the bass too muddy?

The mix engineer or composer who works in surround is not immune to such questions, and in fact, working in surround instantiates an immense, additional set of aesthetic considerations and choices. These include, but are of course not limited to, an expanded questioning of the purpose of the loudspeaker during playback of the mix; the portrayal, creation, or deconstruction of acoustic reality; questions of auditory perspective; and the overriding notion of space as a compositional and mixing parameter.

The role of the loudspeaker. We spoke earlier about the continual self discourse regarding the role of the loudspeaker as musical instrument or sonic window when mixing for stereo. The same question is of course present with mixing for surround, but if anything it becomes an increasingly important question to ask (and perhaps more difficult to answer, as a broader spectrum of possibilities becomes manifest). An understanding of the specific role that the collection of loudspeakers—as well as each individual one—is playing can make the perceptual difference between surround audio as aural gimmick and surround as immersive experience.

The musical "role" that a loudspeaker plays in a mix may seem at first like a silly question to ask. Say we have multitrack-recorded ("tracked," in the parlance) a five-member rock band in the studio. Consider the different effect that assigning each instrument its own discrete loudspeaker in a surround mix would have versus using the many phantom-image possibilities afforded by a surround arrangement to create a manipulated and molded acoustical perspective onto the performance. The aesthetic analog of the first scenario is that the band is in one's living room upon playback: The loudspeakers not only portray individual instruments, but in fact *are* those instruments. This is, in a sense, the aural analog of realist photography, in which the intent of the photographer is that the captured image not only represents but literally is the object of its portrayal.

In the second scenario, the loudspeakers are used as a kind of aural "looking glass" onto a molded and mixed performance. The speakers are not used as instruments, but as a kind of canvas onto which an acoustic space can be created and manipulated. In this case, the locations of the individual loudspeakers are generally not intended to be perceived by the listener: Rather, the mix engineer wants to create a specific acoustical environment that should be perceived in its entirety.

A practical question that arises here is the role of the individual loudspeakers in surround configurations. In 5.1-channel surround, for instance, the center channel is generally used for dialogue and vocal lines, the surrounds are often reserved for "special effects" and reverberant ambience, and the subwoofer is generally reserved for low-frequency rumble effects and added "punch" when necessary. Ultimately, however, there are no hard and fast rules. Trial and error—and lots of patience—are again the best teachers of technique.

Acoustic reality. But at the core of all aesthetic considerations that accompany surround mixing is the portrayal, creation, or deconstruction of a synthetic acoustic reality. Surround audio can bring a much broader palette—a greater potential—to create and alter acoustic environments. A trivial example might be this: Do we want to mix so that it sounds like the listener is sitting literally inside the band or symphony orchestra, or do we want the mix to sound like the listener is sitting further away in the hall? This can be thought of as the question of inside-out versus outside-in perspective.

Related to this idea is the literal inversion of the idea of surrounding a listener with audio externally. If the intent of surround audio is literally to envelop the audience from the outside in, perhaps the inversion of surround is to use spherically radiating loudspeakers to produce sound, just as acoustic instruments radially emit acoustic energy in 360°-space. We return to this notion later in the chapter.

Space as a musical parameter. Another central question is this: Should the chosen aural perspective remain constant throughout the mix, or should it change? Changes of acoustical space tend to occur less frequently in rock music, for example, when it is generally desirable to impart a sense of hearing the band live in a studio or perhaps in some kind of fixed performance venue. In other kinds of music—and especially audio for film and movies—acoustic environments can constantly change. One must continually consider the mechanisms by which acoustical spaces can be changed, whether gradually, abruptly, or somewhere in between.

But space can do more than serve the creation of acoustical environments: It can contribute to a mix or composition as a fundamental musical parameter. Besides helping the listener in navigating acoustical spaces, a mix can allow individual sounds to be panned and moved dynamically. Just as pitch, rhythm, and timbre have historically formed the basis, to varying degrees, for the creation of music, so too now can space join in—thanks largely to resources afforded by the DAW.

To what extent can and should individual sounds move in the constructed space? It might be helpful to consider the fundamental kinds of methodologies in which space can serve as a mixing and compositional parameter. They include space as gestural trajectory, spatial rotation, expansion and contraction, and perspective shift.

TRAJECTORY. Perhaps the simplest spatial archetype in music is antiphony, or "call and response," which dates back thousands of years. Antiphony can be considered a subset of trajectory, for in calling and responding (as with two displaced choirs of singers), an imaginary vector is drawn in space. Surround mixing on the DAW opens an entirely new level of trajectory as a basic musical archetype. By increasing the number of playback loudspeakers required to "perform" a mix, spatial trajectories can ultimately be constructed from anywhere to anywhere. By itself, a trajectory is the movement of a sound source or sources across an imaginary curve in space, and without a change in rotation, size, or perspective of the listening field itself.

ROTATION. With surround audio, the entire sound stage can remain intact while undergoing rotation. Rotating the mix, for example in a clockwise circular motion, can either impart a sense that the listener is remaining still while the imagined sound space is moving, or that the sound space is remaining still while the listener is twirling around. (Think of looking up at clouds in the sky while spinning around.)

EXPANSION/CONTRACTION. In a stereo mix, the sound stage can be expanded or contracted, but generally only by "widening" or "narrowing" the two-dimensional stage. With surround audio on the DAW, the entire three-dimensional listening field can shrink or grow.

CHANGE OF PERSPECTIVE. Yet another way in which space can serve as a compositional parameter is by using it to change the listener's perspective. This is different from changing an acoustical space; with a change of perspective, the band is still playing in the same hall, for example, but the listener is transported around the hall. Or perhaps the listener is brought in for a closer aural "look" every now and then.

The aesthetic considerations involved in creating a surround mix are many. So too are the technical considerations. We now get our feet wet in the technical side by first setting up a project surround studio and then examining the major surround audio formats and channel configurations that can be used.

The Surround DAW Studio

Almost any room that has been properly acoustically treated and is quiet enough can serve as a space for a DAW-based surround mixing studio. It is, however, difficult to design professional-quality studios that work equally well for both surround mixing and stereo mixing, but this is typically less of a concern for project studios. Ideally, the room's length, width, and height dimensions should not form tidy integer ratios (like a 10' × 10' × 10' room, which of course exhibits a *length:width:height* ratio of 1:1:1), but rather they should be as arithmetically "unrelated" as possible to help prevent standing wave interactions and build-up of the eigenmodes. (For example, 14' × 19' × 10' room is often cited as a good small-size studio dimension.) More information on designing room dimensions for studios, and even specifically for surround studios, is available in acoustics reference books.

But short of building or physically altering a room, the chosen space should of course be large enough to accommodate the number of monitor speakers you intend to use. Most studio designers tend to agree that the mixing position point of focus in a surround studio should be facing one of the shorter walls. That is, the median plane of the studio should lie along the longest room dimension, as shown in Figure 7.4, so that the room "appears" long rather than shallow.

Figure 7.4
The ideal median plane in a surround studio.

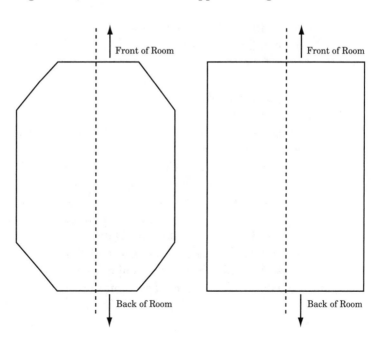

Once the studio is acoustically treated (and perhaps tonally corrected using an permanently installed equalizer, if possible, to tune each loudspeaker to the room), integrating a DAW into the environment can be as simple as attaching the output of the audio interface to powered ("active") monitor loudspeakers or to a surround amplifier (even a consumer receiver with discrete inputs can work) and passive loudspeakers. Most project studios, however, tend to insert a multibus mixer, either analog or digital, between the audio interface and the monitor loudspeakers. Doing so offers two advantages: It allows other instruments and voices to be "patched" into the DAW for hard-disk recording, and it allows the loudness of surround playback to be adjusted during mixing. (Remember to not exceed 85 dB or so, especially for long periods, as this can damage your hearing.)

Most modern surround studios are configured for one surround mixing configuration alone: 5.1. However, others formats are available, most of which were historically developed to accompany commercial film releases, and we talk briefly about each for completeness and some historical perspective.

Introduction to Surround Configurations and Formats

Many different surround channel configurations are available, and each can present its own advantages and disadvantages. These channel configurations can be categorized by their number and suggested arrangement of loudspeakers or the commercial encoding formats that embrace them (for example, Dolby Digital AC-3 uses five full-range loudspeakers plus a subwoofer in a typical 5.1-channel arrangement). Note that the term *configurations* applies to the number, placements, and often the types of individual loudspeakers. The term *format*, on the other hand, generally denotes a company's proprietary encoding/decoding process for surround storage and playback based on one or more configurations. Some configurations, like 5.1, can be encoded in several commercial formats, like Dolby Digital and DTS. And some formats, like Dolby Digital, can represent audio in different configurations, like 5.1, LCRS, and 2.0.

We will talk about each major surround configuration, including suggested loudspeaker placements for mixing and playback. The discussion of each configuration includes a listing of the supported commercial formats.

Quadraphonic   Although earlier attempts were made to enhance the realism afforded by the new stereo recording and playback techniques, such as playing reverberant images in loudspeakers placed in the back of auditoriums, the first commercially viable (well, not really, but we'll get to that) surround configuration is the quadraphonic arrangement. Three-channel sound reproduction systems (using front left, front center, and front right channels) had been used occasionally in the 1930s, for example, by the conductor Leopold Stokowski and the Philadelphia Orchestra for recording and transmission to another remote concert hall, as well as in early experiments in Bell Laboratories, but over thirty years passed before the notion of "surround" encroached upon the common audio parlance. (Another famous example is "Fantasound," a three-channel LCR configuration that Disney employed for the 1940 move *Fantasia*.) Although no one standard encoding format ever emerged as a winner, a standard loudspeaker configuration did: The four speakers were simply spaced 90° apart in the four corners of an imaginary square, as shown in Figure 7.5.

Figure 7.5
Quadraphonic
loudspeaker
arrangements.

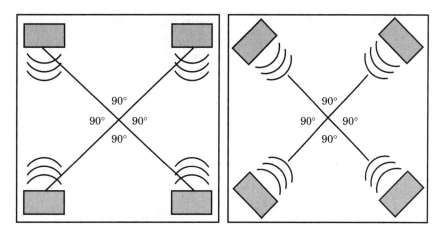

In 1970, the first quadraphonic recording was issued. Quadraphonic sound ultimately failed in the commercial music marketplace, partly because no standard encoding format emerged; a quadraphonic mix could sound drastically different if played back using another company's decoding equipment rather than the one used to make the mix. It also became clear that basic psychoacoustic principles were not taken into account. (For example, placing two loudspeakers in front of the listener and two behind seems to imply equivalence in frontal and rearward hearing. On the contrary, the ears are angled slightly forward, helping to create a more sensitive frontal listening focus.) And, for various reasons, consumers at the time were less likely to purchase the additional loudspeakers required to play quadraphonic recordings. Nevertheless, quadraphonic took hold with many electronic music composers who enjoyed the compositional possibilities afforded by an equal spatial distribution of loudspeakers, and this equal loudspeaker distribution ultimately gave way to the more popular octophonic arrangement commonly used today in the computer music community.

Incidentally, at least part of the interest in four-channel sound can be traced to the earlier work of Jacques Poullin, a French engineer who developed a four-channel sound playback system in 1951 for composers Pierre Schaeffer and Pierre Henry. The system used two frontal, one rear, and one overhead loudspeaker placed in a concert hall, and sound could be distributed to them with an instrument called the *Pupitre d'Espace* ("Space Console"). By waving one's arms around in space, audio would be diffused around the concert hall in gestural tandem with the operator's arms.

LRS and LCRS surround. Dolby Laboratories (http://www.dolby. com) introduced Dolby Surround in 1982, which called for front-left, front-right, and surround speakers. However, the surround speaker(s) (as few as two but typically more, particularly in movie theaters) all played the same audio; this is referred to as an LRS (left-right-surround) system. Dolby Surround called for LRS audio to be stored on only two channels, which were then decoded on playback into three (Figure 7.6a).

Figure 7.6
(a) LRS; (b) LCRS

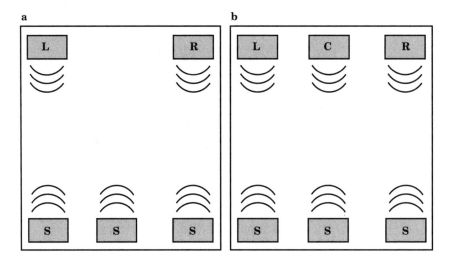

As previously mentioned, three-channel stereo reproduction dates back to the 1930s. The addition of a center channel fills in the gaps between the left and right channels when their corresponding loudspeakers are placed too far apart to make phantom imaging very effective (for example, across a movie screen, or on either side of a large stage).

The addition of a center channel to LRS Dolby Stereo led to another matrixed surround format known as Dolby ProLogic. This format was developed for film soundtracks and not specifically for music, although some have encoded music in ProLogic. Dolby ProLogic allows LCRS audio to be stored on an analog two-channel stream, decoded later into any of several channel configurations, up to and including LCRS (Figure 7.6b).

5.1-channel surround. The next step in commercial surround configurations was a full-blown five-channel surround configuration, enhanced by a subwoofer. The 5.1 (pronounced "five-point-one") configuration, perhaps the most commonly used surround format today, includes front-left (L), front-center (C), front-right (R), left-surround (Ls), and right-surround (Rs) channels, plus a subwoofer, also called the LFE, or "low-frequency effects," channel. The ".1" refers to the fact that subwoofers can only reproduce very low frequencies, and so the bandwidth allocated to a typical full-range audio channel is not required to encode the LFE channel. We can get by with a much smaller bandwidth, or frequency range, thereby diminishing the amount of bits needed to store 5.1-channel audio.

Actually, the bandwidth allotted to the LFE channel is only 120 Hz, which is about the highest frequency most subwoofers are used to produce. This means that the LFE channel can be stored at 1/200 of the full-channel 48-kHz sampling rate to save space. (Recall that the sampling rate must be double the highest representable frequency; 48,000 Hz ÷ 200 = 240 Hz, which can represent up to 120 Hz.) But, as Tomlinson Holman (who coined the term "5.1" in 1987) notes, "5.005" just isn't as snazzy as "5.1"!

Unlike the case with quadraphonic sound in the 1970s, a standardized specification for 5.1 channel loudspeaker placement and calibration has emerged. The International Telecommunications Union's specification ITU-R BS.775 calls for surround loudspeakers to be placed in a prescribed way, shown in Figure 7.7. The center speaker is placed directly in front of the listener, while the front-left and front-right speakers are displaced 30° off-center, about 2 to 4 meters

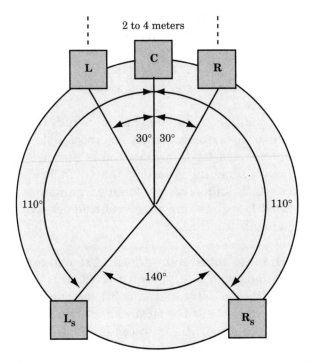

apart. The two surrounds should be placed 110° ± 10° (i.e., 100° to 120°) relative to the front-center location.

All five speakers should be placed the same distance from the mixing position. Furthermore, it is suggested that the front-left loudspeaker, front-right loudspeaker, and the mixing position form the vertices of an equilateral triangle.

Two loudspeakers can be somewhat tricky to place: the center channel and the subwoofer. Because the DAW is based around a computer, a monitor is necessarily required, but its placement can interfere with the center channel. Some studio furniture, for example by Omnirax (http://www.omnirax.com) or Raxxess (http://www.raxxess.com), is specifically built to accommodate both a computer monitor and a center-channel loudspeaker. It is often convenient to place the center channel just above the computer monitor. Another alternative is to purchase a glass-top computer desk that includes a recessed under-mount support for the computer monitor. With such a computer table, the computer monitor is placed entirely underneath the desk, out of the way of the center channel.

There is no standard for subwoofer placement; just keep in mind that placing a subwoofer (or any speaker for that matter) in the corner of a room yields a very noticeable boost in the bass response, so equalization may be necessary. In general, any placement from slightly off-center behind the center channel to just outside the imaginary circle enclosing the five primary loudspeakers can work, depending on the room acoustics.

Once the speakers and subwoofer have been properly placed in the room, their levels should be calibrated. The typical reference level used for the main loudspeakers in most project studios—whether stereo or surround—is 85 dB SPL. This means that, with all faders on the DAW set to an output level of 0 dB (including the faders on any mixer to which the DAW is attached), playing a pink noise sound file should cause a sound level meter to read 85 dB. The subwoofer is typically calibrated to a reference level of about 4 dB higher relative to the main loudspeaker's reference level. Reference levels vary from studio to studio and depend on the mix engineer's taste and the nature of the material being mixed (for example, whether it is audio for film, classical music, or heavy metal).

Thanks to standard 5.1 formats and configurations, a great many tools are readily available for the DAW when mixing and mastering in 5.1 surround. And budget DAW surround studios can even use a consumer/prosumer home theater amplifier to drive the five main loudspeakers. In fact, creating a project 5.1 DAW-based studio need not be much more expensive than its stereo counterpart.

There are two primary formats used to compress and encode 5.1-channel audio: Dolby Digital and DTS. Both systems are lossy. The first such format, Dolby Digital SR•D, was developed in 1992. Intended solely for use in film sound, it enabled discrete 5.1-channel audio to be optically encoded onto 35 mm film between the film sprocket holes. The consumer "home" version of Dolby Digital 5.1 is called AC-3, which incidentally has been adopted as the future surround distribution standard in the United States for high-definition television broadcasts. One of the nice features of Dolby Digital is that DAW plug-ins are available to encode 5.1-channel mixes in this format.

Another competing format for encoding and decoding 5.1 discrete audio channels is called DTS from a company of the same name (http://www.dtsonline.com). The DTS format was first used in 1992 to encode the surround soundtrack for Stephen Spielberg's movie *Jurassic Park*. DTS can encode data at a higher bit rate than can Dolby Digital, but because their encoding algorithms are fundamentally different, this does not necessarily translate into better sound quality. There is actually much debate in the industry over which sounds "better," and there does not seem to be a clear winner yet.

Bass management.   A term often encountered in surround audio—not exclusive to 5.1—is *bass management*. Surround audio can be mixed on a DAW either with or without a bass management system. Without it, the source channels (L, C, R, Ls, Rs, and LFE) are all treated and routed discretely and independently (see Figure 7.8a). In other words, all the designed output busses from the multitrack editor-sequencer correspond directly to the output channels themselves, with no overlap.

Figure 7.8
Signal flow charts of (a) 5.1 mixing without bass management; (b) 5.1 mixing with a typical bass-management system.

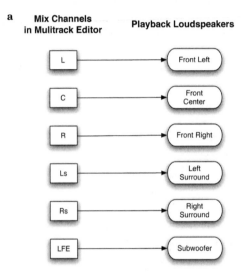

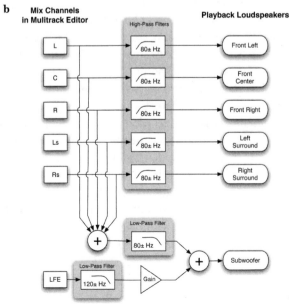

Using a bass-management system (of which many are available), on the other hand, reroutes low-bass frequencies from each of the main channels (L, C, R, Ls, and Rs) to the subwoofer, adding them together with the low-frequency audio information that the LFE channel already contains. The LFE channel is low-pass filtered to ensure it isn't asked to produce frequencies that it can't, and each of the main channels are conversely high-pass filtered.

Most DAW multitrack mixers include bass-management functionality, enabling the user to define the precise parameters of the system. Furthermore, plug-ins can provide additional bass-management tools.

6.1-channel surround.   The addition of a single center surround (Cs) results in a 6.1-channel configuration: L, C, R, Ls, Rs, Cs, and LFE. And as with 5.1, two primary formats are used to encode and distribute 6.1-channel audio: Dolby Digital EX and DTS-ES Extended Surround. Both of these formats employ matrix decoding of the three surround channels.

7.1-channel surround.   The earliest record of a discrete seven-channel system is Cinerama, which was developed by inventor Fred Waller and premiered in 1952. Cinerama, a kind of precursor to modern-day IMAX theaters, used not only seven discrete audio channels but also employed three separate film strips placed side-by-side inside the playback projector to create a truly widescreen image on a giant curved screen.[2] Cinerama soundtracks were magnetically recorded onto a separate 35 mm film strip and included left, left-center, center, right-center, and right channels for playback on stage. Two channels of surround audio could be dynamically routed to the rear wall, left wall, right wall, or any combination thereof during film playback. Unfortunately, Cinerama only lasted just over a decade.

The 7.1 configuration calls for two front speakers to be added to the standard ITU 5.1-surround arrangement. These speakers, called left-center (Lc) and right-center (Rc), are placed between the left and center speakers and the right and center channels, respectively, as shown in Figure 7.9. This yields a surround configuration of L, Lc, C, Rc, R, Ls, Rs, and LFE.

The standard format used with discrete 7.1-channel surround is Sony Dynamic Digital Sound (SDDS), for which many film sound-

---

[2] *This is Cinerama*, a plotless quasi-documentary, was the highest-grossing film of 1952!

Figure 7.9
Suggested 7.1-
channel
loudspeaker
arrangement.

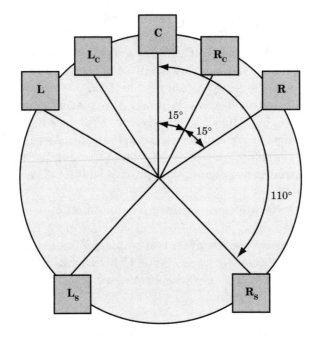

tracks are often mixed. The SDDS format compresses audio using a proprietary algorithm, yielding a data reduction rate of about 5:1. Lexicon's Logic 7 format is a matrix-decoded specification for 7.1-channel surround, although the recommended speaker locations are spread out around the front more, such that two of the loudspeakers are more or less directly to the side of the listener.

8.0-channel surround. Octophonic, or 8.0-channel, arrangements of loudspeakers have been common in the electronic and computer music community for the past several decades. Octophonic audio developed for several reasons among composers, largely owing to the equal coverage it provides to all listening angles as well as the ease of playing back octophonic mixes from 8-track analog tape or, more recently, a low-cost digital multitrack recorder, like the Alesis ADAT. The first known octophonic music was composed by John Cage; his Williams Mix (1951–1953) was mixed onto eight separate quarter-inch magnetic tapes to be played back simultaneously. Many other works have followed in the eight-channel format up to the present.

Two octophonic loudspeaker arrangements are commonly employed. The first, shown in Figure 7.10a, involves four pairs of loudspeakers equally distributed around the audience: front left, front right, front side left, front side right, rear side left, rear side right, surround left, and surround right. By rotating the sound field 22.5°, we arrive at the arrangement shown in Figure 7.10b. This arrangement involves three pairs of loudspeakers with a front center and rear center: front left, center, front right, side left, side right, surround left, surround center, and surround right.

Figure 7.10
(a) One commonly used octophonic loudspeaker configuration; this arrangement was used by composer Iannis Xenakis for his composition Bohor (1962); (b) another arrangement occasionally encountered; many other arrangements are routinely used in the electroacoustic music community.

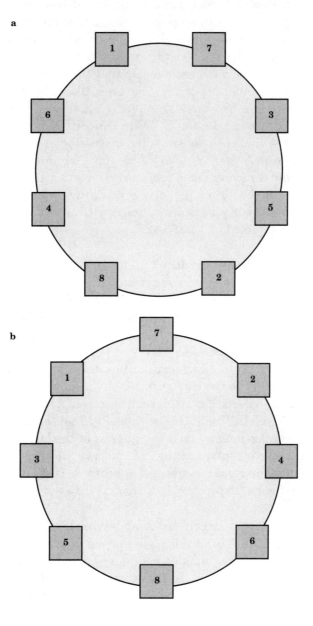

Octophonic mixing environments are perhaps most commonly found in academic computer music and composition studios. One might argue that such an arrangement tends to ignore the basic psychoacoustic features of human listeners in that we are more sensitive to what happens in front of us than in back of us, and that equally distributing the sound reproduction forces around a circle disregards this fact. However, octophonic arrangements can be very effective musically, particularly in large concert halls, where composers might wish to create antiphony from all possible angles with respect to the audience.

One of the convenient features of mixing and playback with eight channels, as we mentioned, is that the entire mix fits perfectly on an eight-channel MDM recorder tape, like a DA-88 or ADAT. However, composers are more frequently playing back multichannel music direct from a laptop DAW in concert (particularly live-electronics pieces that involve computer processing of a live performer), and so an even larger number of playback channels are easily available. Many composers also seem to be adopting the commercial formats of 5.1 and 7.1 for mixing, primarily owing to the wealth of available surround mixing software coupled with new multichannel distribution formats like DVD and SACD.

10.2-channel surround.   The 10.2-channel configuration, proposed by Tomlinson Holman, augments the surround capabilities of the standard 5.1-channel arrangement. Although no standard for 10.2-channel surround has yet emerged, Mr. Holman specifies that to the 5.1-channel configuration should be added a rear-center loudspeaker, left and right loudspeakers at ±60° and two front "height" channels placed at ±45° but raised higher than the others as shown in Figure 7.11. The ".2" refers to the configuration's two subwoofers, which allow low-frequency-effects spatial placement in a 10.2 mix to some extent.

Some DAW software, such as Pro Tools and Digital Performer, support 10.2-channel mixing right out of the box with panning tools and extensive signal-routing capabilities. However, with an appropriate audio interface (or several), almost any DAW multitrack mixer can be configured to produce 10.2-channel (or even more) surround music.

16-channel surround and beyond.   Phantom images can be a wonderfully successful aural illusion: If done correctly, we can actually be tricked into thinking a sound source is located somewhere between two loudspeakers. But why not simply place another loudspeaker where we want the phantom image to occur? Indeed, the "brute force"

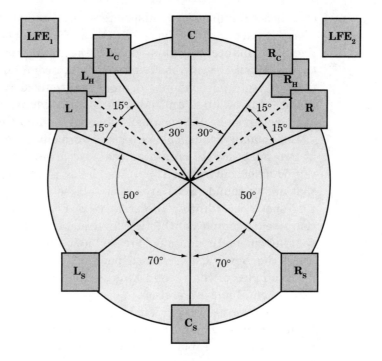

Figure 7.11
Onesuggested
10.2-channel
configuration.

to sound spatialization, in which maybe 16, 24, or even hundreds of loudspeakers are used to play back sound, has the potential to portray an immersive audio environment of a much higher quality, and with a greater sense of realism, than phantom images could ever create. Of course, with additional loudspeakers comes additional expense (and few people wish to look at more than a few loudspeakers in their living rooms). Furthermore, few performance venues are equipped to support such an approach to surround audio playback.

Nevertheless, the "brute force" approach to surround audio has received some attention in the past, primarily with avant-garde electronic musicians as well as in theatrical productions. Very little music is typically mixed into 16 or more discrete channels; beyond discrete octophonic or 10.2-channel mixes, it is much more common to find music mixed to few channels "upmixed" or "diffused" throughout a large space during playback. We talk more about this later in this chapter.

Part of the reason for not mixing in a very large number of channels is the historical difficulty and expense of storing and playing back so many channels. The Italian Futurist movement in the early part of the twentieth century played different recordings and amplified sounds over large numbers of loudspeakers, and later composers, such as Pierre Schaeffer, John Cage, and David Tudor constructed music and live performance situations comprising large numbers of playback channels, but the playback channels often came directlyin real time from discrete microphones or nonsynchronized record players and radios. This tradition continued with multichannel real-time performance instruments, such as the 24-channel Sal-Mar Construction.

But so many loudspeakers have historically been used to serve one of two purposes: to create as realistic a multidimensional sound-reproduction environment as possible; or to create a playback space that allows sounds to be arbitrarily placed and moved around in virtual space according to the creative desires of the mix engineer or composer. Examples of the first kind include some of the early multichannel cinematic and music-reproduction formats that attempted to assist the viewer/listener in suspending disbelief, at least temporarily. Examples of the second kind include elements drawn from the experimental music tradition.

But it seems perhaps that indeed, all roads do lead to Rome. The technical, creative, and commercial, and marketplace forces all definitely seem to be rapidly embracing multichannel audio, leading, one hopes, to at least some kinds of standardization along the way.

What about MPEG?   Recall from Chapter 3 that recent incarnations of the MPEG audio specifications support multichannel audio. Although MPEG-2 Advanced Audio Coding (AAC) and MPEG-4 already support the encoding of multichannel audio (and the MPEG-7 and MPEG-21 proposals indicate these new standards will as well), they have not been used as widely in the music industry for distribution and delivery of surround content, at least at the present.

A note on downmixing.   Because all environments can support surround playback, downmixing of a surround mix is often used for certain delivery formats, depending on the target venue or audience. Many commercial formats, like Dolby Digital and DTS, include built-in procedures for automatically downmixing a mix to two channels when only stereo playback is available. How does one compact a surround mix into stereo without messing up everything? There is no one answer; many different techniques exist. If you are mixing for a commercial format, it is necessary to check the mix after encoding on both surround and stereo playback equipment.

Even if a surround mix is not intended for commercial distribution, it is a good idea to occasionally check it in stereo, just as it is a good idea to check stereo mixes in mono. This can assist in locating phase cancellations, and in particular it can ensure the mix will sound as expected in the event surround playback is not possible. Because 5.1 is the most commonly used surround configuration, it has been well studied, and several algorithms have been proposed for downmixing 5.1 surround mixes to stereo.

The Dolby Digital format itself specifies its own internal downmixing parameters. Once a surround mix has been encoded into a Dolby Digital bitstream, it can then be decoded and played in mono, stereo, four-channel surround, or in the original 5.1 configuration according to the decoding equipment used and the preference of the listener.

Another option for downmixing is to encode surround content into a "virtual" surround format such as SRS Circle Surround (http://www.srslabs.com). Using a Circle Surround encoding plug-in, a discrete, multichannel surround mix can be downmixed to stereo; when the stereo mix is played back over SRS-compliant equipment (like a home-theater decoder/amplifier) connected to only two loudspeakers, a "virtual" surround playback environment is created.

The inside out.   Our discussion of surround audio so far has been predicated entirely on the notion that sound waves radiate from an external source surrounding the listener (like a 5.1-channel speaker array). And clearly, sound must be emitted from an external location inward to literally surround the listener, by definition. But short of bone conduction, a process by which bones in the skull act as mechanical transducers to help us hear sound, another approach is available: spherical loudspeakers. By radiating sound from the inside-out rather than emitting it directly to the listener from the outside-in, speakers that are designed to radiate sound in 360 degrees can accomplish two primary features: (1) they can emulate the actual three-dimensional radiation patterns of any acoustic object, like a guitar or violin; and (2) they can couple directly in a room, working in tandem with the inherent acoustics of the room to achieve surround sound envelopment.

Spherical and other alternative loudspeakers are an active ongoing research area, and many prototype systems rely heavily on the digital-signal processing horsepower of the DAW to assist them in their quest to reproduce sound as faithfully and naturally as possible.

Ambisonics. One of the most unique approaches to multichannel, surround-sound recording, playback, synthesis, and mixing, is embodied in a system known as *ambisonics*. One of the ideas behind ambisonics is the idea that hundreds or even thousands of point-source playback loudspeakers would be required to faithfully reproduce virtual three-dimensional space using phantom images and panning.

First proposed in the early 1970s by Michael Gerzon, ambisonics is a system whereby three-dimensional audio information is captured with a special microphone called a *soundfield* microphone. Soundfield microphones embody four diaphragms (called W, X, Y, and Z) arranged in a spherical configuration. The X channel represents the front-back dimension ("depth"), the Y channel represents the left-right dimension ("width"), the Z channel represents the up-down dimension ("height"), and the W channel represents the omnidirectional sound pressure level to which the other channels are all referenced. Once the audio has been recorded into a four-channel stream (called B-format recording), the original three-dimensional sound space can be mathematically reconstituted and reproduced using any number of output loudspeakers.

One clear advantage of ambisonics is the system's decoupling of recording from reproduction. Another is the sheer and remarkable fidelity with which three-dimensional sound spaces are reproduced. However, a limiting factor for many is cost: A complete system, including soundfield microphone, encoder, and decoder, can be quite expensive.

Spatial-Processing Techniques

Sounds can be artificially placed in three-dimensional virtual space within a mix using any of various signal-processing techniques. These include panning, manipulating the ear's timing and level cues, reverberating, filtering, using ambisonics processors, and manipulating the ear's tonal cues by imposing a set of head-related transfer functions onto a sound file.

Panning and Equal-Power Laws

Perhaps the simplest way to place sounds in a sound field when mixing, whether in stereo or a multichannel configuration, is by manipulating the ILD cue. For instance, to place a mono sound file to the left in a stereo mix, one could send copies of the sound file to both the left and right output channels and simply attenuate the amplitude on the right channel. The louder the left channel is perceived relative to the right—that is, the greater the interaural level difference—the stronger the ILD cue that tells the brain to consider the sound as coming from the left (provided the frequencies are high enough for the ILD to serve as a salient spatial-perception cue).

A device that automates this process of ILD manipulation is called a *panner*, of which the most common type is a stereo panner. The only thing really necessary to understand about panning is the fact that humans respond more to the power of acoustical signals than directly to their amplitudes regarding spatial perception.

Consider a mono sound file that we wish to pan from left to right across the sound stage. We could simply time-align copies of the corresponding sound file on the left and right channels, fading out the left-channel's file as we fade in the right-channel's file. (See Figure 7.12.)

If our fade curves are linear (straight-line) in nature, however, we will perceive a "hole" in the middle; that is, the sound will sound quieter, and hence farther away, in the middle of the fade. This is a direct result of the fact that we respond here to power and not amplitude: In fact, the perceived intensity of two or more sounds played together is proportional to the square root of the sum of the squares of the individual amplitudes:

$$\text{Intensity} \propto \sqrt{a_1^2 + a_2^2} \text{ for two point sources} \tag{1}$$

$$\text{Intensity} \propto \sqrt{\sum_i a_i^2} \text{ for } i \text{ point sources} \tag{2}$$

This means that, at the center of the linear fade—where the amplitude of each sound file is reduced by 50 percent, the resulting perceived intensity would be about $\sqrt{(0.5^2) + (0.5^2)} \approx 0.707$, which is about 30 percent less than the perceived intensity at the extrema of the pan.

To correct this, DAW software employs *equal-power panners*, just as hardware console panners do. Instead of linear fades, equal power panners use logarithmically tapered fades chosen so that no "holes" in the sound are created. Consider the resulting intensity of a phantom image panned to the center, in which each logarithmic fade curve has a value of $\frac{\sqrt{2}}{2}$:

Figure 7.12
(a) Equal-power
panning fades; (b)
linear panning
fades.

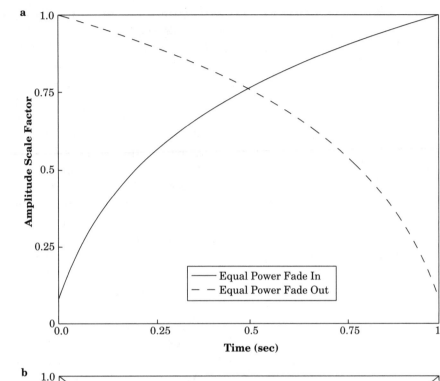

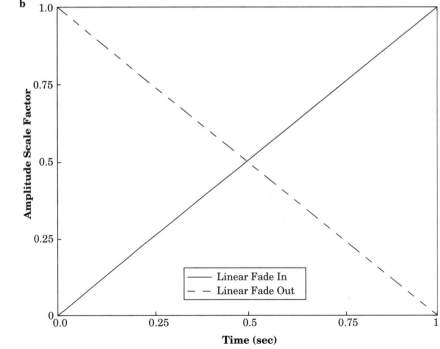

$$\sqrt{a_1{}^2 + a_2{}^2} = \sqrt{\left(\tfrac{\sqrt{2}}{2}\right)^2 + \left(\tfrac{\sqrt{2}}{2}\right)^2} = \sqrt{\tfrac{2}{4} + \tfrac{2}{4}} = 1$$

which is what we want.

The panning problem is confounded when working in surround, but thankfully many off-the-shelf DAW tools are available to crunch the numbers behind the scenes. In particular, several new dimensions of control are opened in surround:

- How many speakers should be included in a pan when creating phantom images? That is, what image width is desired?
- Sounds must be able to be placed outside (easy) or inside (difficult) the imaginary circle on which the loudspeakers lie.
- A simple knob can no longer be used. Does a joystick, trackball, mouse, or something else provide the optimal control solution?

Fortunately, commercially available panning plug-ins address these issues; some even allow a virtual panning "puck" to be controlled via virtually any universal serial bus (USB)-compliant hardware controller.

ITD

Another way to place sounds in virtual space is to manipulate or artificially create interaural timing differences, particularly with lower-frequency sounds. Even without changing the amplitudes of any of a mix's output channels, short timing delays (in the millisecond range) can impart very strong spatial cues.

Reverberation

Both ILD and ITD manipulation form the basis of many left-to-right or circular panning algorithms. But careful use of reverberation can impart a spaciousness of dimensionality to a sound file that can simulate distance from the listener quite effectively. Background tracks or more ambient sounds meant to be perceived as farther away are often processed using reverb plug-ins. But be careful again about the implications, intentional and especially unintentional, of placing sound sources in the same mix within different physical spaces.

Filtering

As a corollary to reverb, low-pass filtering of a sound source can also make it sound farther away. In fact, virtually all reverb plug-ins and algorithms incorporate some sort of low-pass filtering, a natural result that occurs when sounds are physically displaced far away from the listener.

Ambisonics

Ambisonic techniques can also be used to place sounds in virtual, three-dimensional spaces in a mix. In fact, several free ambisonics spatial-processing plug-ins, authored by ambisonics proponents Dave Malham and Ambrose Field, are available from the Music Technology Group at the University of York (http://www.york.ac.uk/inst/mustech).

HRTF Processing

Finally, head-related transfer functions can also be used to create virtual space in a mix. By using an HRTF-based spatial-processing plug-ins (also called *binaural* processing plug-ins, because they attempt to mimic how our two ears work together to localize sounds), sounds can be placed in virtual three-dimensional space, although the technique is generally useful primarily for headphone playback of a mix. Owing to the sensitivity and variability of HRTF data sets, though, the effectiveness of this technique can vary depending on the listener and how close the data set used matches the physiological traits of the actual listener.

Wave-Field Synthesis

The "brute-force" approach to the presentation of realistic three-dimensional audio has most recently been manifest in a suite of technologies known as wave-field synthesis (WFS), the subject of much current research. (See Figure 7.13.) Using this technique, large arrays of many loudspeakers are driven by many computers to literally synthesize the components of each wavefront emitted by a sound source. According to *Huygens' principle*, a natural acoustical wave-

Figure 7.13
A wave-field
synthesis
installation at the
Friedrich-
Alexander
University,
Erlangen-
Nürnberg,
Germany.

Figure 7.13
A wave-field
synthesis
installation at the
Friedrich-
Alexander
University,
Erlangen-
Nürnberg,
Germany.

front emitted from a resonating body cannot be distinguished from one created with an infinite number of point sources along each point of the wavefront. WFS creatively uses convolution and DAW processing to harness this principle, allowing sounds to be virtually placed with great three-dimensional precision in a listening environment. In addition to the immense sense of realism that can be created with the WFS technique, the listening "sweet spot" is also much larger than is the case with other multichannel presentation techniques.

Distribution and Presentation of Surround Audio

Surround audio is literally all around us—pardon the pun!—in consumer electronics, movie theaters, new cars, and concert halls. And with the rapidly diminishing price of creating a surround home theater, more and more people are able to routinely listen to audio in a surround environment.

As surround audio continues to mature both artistically and commercially, mix engineers and composers must be armed with a working knowledge of the ways in which it can be distributed and presented. To conclude this chapter, we first summarize the primary distribution formats for surround audio, and then we survey ways in which surround audio can be presented to listeners.

Distribution of Surround Audio

Surround mixes can be presented and distributed in many ways: on a DVD, on a multichannel tape, or over the Internet, as we have already discussed. They can even be played back directly from a DAW into a surround loudspeaker arrangement in a performance space.

Commercial formats can be characterized by whether they employ discrete channels (in which each audio channel remains fully intact and independent of the others throughout the encoding and decoding), or computed channels (with one of several processes collectively called *matrixing*, in which some output channels may be analytically computed and derived from others). We can also characterize commercial surround audio formats in terms of the distribution media that support them.

Rather than include a thorough discussion here of distribution formats for surround audio, we include one in the final chapter of this book, which is exclusively devoted to the topic of mastering and distribution. (Feel free to skip ahead if you ae so inclined.)

Presentation of Surround Audio

We are all probably most familiar with the presentation of surround audio in the movies. And thanks largely to the commercial success of the DVD format, an increasing number of people are able to affordably enjoy surround audio at home.

The art music world has also enjoyed a long tradition of presenting multichannel audio to audiences in concert halls and various performance venues. Short of designing custom controllers like the *Pupitre d'Espace* of the 1940s for real-time "diffusion" of an audio channel to a collection of loudspeakers, the traditional sound-reinforcement mixer has served as something of a performance instrument in the concert presentation of surround audio.

Early examples of two-channel or multichannel music that was "upmixed" during performance to a large number of loudspeakers include works by Stockhausen and Varèse. The composer Karlheinz Stockhausen has indeed enjoyed a long-lived involvement with multi-channel sound. His 1956 composition *Gesang der Junglinge* ("Song of the Children") was composed for five-channel magnetic tape distributed around the performance space. As the composer wrote, "The direction and movement of the sounds in space is shaped by the musician, opening up a new dimension in musical experience." Edgard Varèse's classic *Poème Electronique* for two-channel tape, specifically composed for performance at the 1958 World's Fair in Brussels at the Le Corbusier-designed Phillips Pavilion, was played using over 400 separate loudspeakers.

This tradition of sound diffusion in concert has led to the construction of special-purpose concert halls for the performance and diffusion of surround music. Examples of these include the Audium in San Francisco and the Audium in San Francisco (Figure 7.14), among several others.

Figure 7.14
The Audium in San Francisco, California. Photo courtesy Stan Shaff.

Movie theaters and concert halls are not the only performance venues historically associated with surround audio. Planetaria, too, have been involved in high-quality visual and surround music presentations for some time. Beginning in 1957, the Morrison Planetarium in San Francisco presented its Vortex series of concerts, featuring a custom-built 38-channel surround playback system. The series included presentations of surround music composed by Vladimir Ussachevsky, Luciano Berio, Toru Takemitsu, Giorgy Ligeti, and others.

But one particular problem of presenting surround audio to an audience—whether in a movie theater, a concert hall, a planetarium, or anywhere else—is the traditionally limited size of the "sweet spot." A technology that we mentioned earlier called wavefield synthesis, promises to widen the listening sweet spot for the presentation of multichannel music in large spaces. In fact, WFS can create virtual sound projections that can be accurately perceived by the majority of listeners in the space. WFS systems, which are capable of simulating an unbounded number of sound point-sources, have already been installed in several venues, including a movie theater in Germany.

Who knows what the future of multichannel audio will be? Many of the technologies involved have certainly existed for quite some time, but economic and cultural factors—not to mention overwhelming marketplace response to multichannel video and audio programming content—indicate it is here to stay this time. And owing to the exponentially increasing number and quality of surround software tools, the modern DAW will certainly play an increasing role in its future.

For Further Study

Blauert J. *Spatial Hearing*, 2nd ed. Cambridge, Massachusetts: MIT Press; 1999. Originally published as *Räumluches Hören*, by S. Herzel Verlag, 1974. Chapters 1–4 translated by John S. Allen. Cambridge, Massachusetts: MIT Press: 1997.

Carlile S. *Virtual Auditory Space: Generation and Applications*. Chapman and Hall: New York; 1996.

Clozier C. "The Gmebaphone Concept and the Cybernéphone Instrument." *Computer Music Journal* 25(4):81–90.

Duda R. "3-D Audio for HCI." http://interface.cipic.ucdavis.edu/CIL_tutorial/3D_home.htm.

Gardner B, Martin K. "HRTF Measurements of a KEMAR Dummy-Head Microphone." http://sound.media.mit.edu/KEMAR.html. MIT Media Lab.

Handel S. *Listening*. Cambridge, Massachusetts: MIT Press; 1989.

Holman T. *5.1 Surround Sound: Up and Running*. New York: Focal Press; 1999.

Kendall G. "A 3-D Sound Primer: Directional Hearing and Stereo Reproduction." *Computer Music Journal* 19(4):23–46.

Malham D. "Toward Reality Equivalence in Spatial Sound Diffusion." *Computer Music Journal* 2001;25(4):31–38.

Malham D. *Sound in Space Home Page.* http://www.york.ac.uk/inst/mustech/3d_audio/welcome.html.

Mamalis N. "The Musical Character of Space." *Proceedings of the Third Triennial ESCOM Conference,* 1997; p. 392–394 (in French).

Rumsey F. *Spatial Audio.* New York: Focal Press; 2001.

"Surround Sound: Past, Present, and Future." http://www.dolby.com/tech/p-pr-fut.html

Terrugi D. "Digitally Produced Music and Hand-Controlled Sound Diffusion Systems: The Acousmonium and the Performance of Acousmatic Music." *Vortrage und Berichte vom KlangArt-Kongress 1995 an der Universitat Osnabruck, Fachbereich Erziehungs- und Kulturwissenschaften,* 1998; p. 383–385 (in German).

Tochimczyk M. "From Circles to Nets: On the Signification of Spatial Sound Imagery in New Music." *Computer Music Journal* 2001;25(4):37–54.

Zvonar R. "A Chronology of Bay Area Multichannel Music and Related Events." http://www.zvonar.com/writing/Surround_the_Bay/sidebar_chronology.html

Exercises and Classroom Discussion

1. Investigate the history behind multichannel audio from the perspectives:
   a. Film music
   b. *Musique Concrète* in France
   c. Elektronische Musik in Germany
2. Why did quadraphonic sound ultimately fail in the 1970s? Can you find any other historical surround attempts that failed in some way?
3. What advantages and disadvantages does the DAW present when working with surround versus a traditional console-based studio?
4. How does space compare to pitch, rhythm, and harmony in the grand musical scheme of things? Why?
5. Listen to music composed for virtual 3-D playback over headphones.
6. Can you think of an example drawn from an art form outside of music that is analogous to our instrument-versus-window role of loudspeakers discussion with surround mixing?
7. What are the advantages and disadvantages of standardizing speaker locations and surround formats?

8. Is the mixer a musical instrument? Can a concert hall be a musical instrument? What is a musical instrument, anyway?

9. Practice diffusing a stereo piece over multiple loudspeakers using a mixer.

10. We spoke briefly about mixing from the "inside out" using spherical speakers. Research the history of alternative loudspeaker designs.

11. What DAW software tools are available for sound diffusion? Practice diffusing a stereo piece over multiple loudspeakers using software and compare your impressions to using a traditional mixer.

12. Create a surround re-mix of a stereo piece you did not write or mix. (You might try ripping a compact disc to get material.)

13. Create a stereo re-mix of a multichannel piece you did not write or mix. (You might try ripping the audio portion of a commercial DVD-Video and decode the Dolby Digital bitstream with a DAW plug-in.)

14. Locate a cliché spatial gesture in a commercial DVD movie. What makes the gesture a cliché, and is it effective? Why or why not?

15. Create a more complete taxonomy of spatial trajectories. How would you arrange it?

16. Spatialize a single sound effect with a binaural plug-in processor, and play the result over headphones to several friends. Do they perceive the sound as occurring in the same three-dimensional place as you intended?

Recording and the DAW

So far, we have taken the pre-existence of sound files on the DAW for granted while learning about the basic physics of sound, hearing, mixing, audio file formats, plug-ins, and the historical context for the modern DAW. But before we can mix on a DAW, we need sound files as the basic building blocks; these can come from three places. They can be ripped from existing recordings (for example, sound effects libraries and sample libraries), they can be synthesized from scratch (for example, with a software synthesizer on a DAW or using a sound synthesis programming language), or they can be obtained from recording a sound source. We've spoken a bit about the first two options (and more will be said in the next chapter on the subject of synthesis), and the third and final option is the subject of this chapter.

The chapter begins with a discussion of the idea of *transduction* and then addresses the specifics of microphone types, terminology, and characteristics. Next, specific microphone placement techniques are discussed for mono, stereo, and multichannel recordings, and the chapter concludes with a general discussion of recording issues that are specific to the DAW.

Transducers

Newtonian mechanics teaches that energy can exist in two fundamental forms: potential and kinetic. Potential energy refers to the ability to do physical work owing to the physical position of an object in a field, such as a gravitational field or an electromagnetic field. Kinetic energy, on the other hand, refers to the ability to do physical work owing to a mass that is in motion.

Potential and kinetic energy come in different flavors that can be manifest in various ways, including voltage (electrical potential energy), a squished-together spring (mechanical potential energy), and a speeding bullet train (mechanical kinetic energy).

A *transducer* is a device that converts energy from one form to another. Although it sounds like science fiction, it's actually quite simple conceptually. Think of what would happen if one placed a sheet of paper in front of a subwoofer, and if in front of the paper were placed a row of dominoes lined up in front of each other. With enough acoustic force from the subwoofer, the paper could physically move enough to push over the dominoes. Two transducers were involved in this thought experiment: the voice coil of the subwoofer, which converts electrical energy (in the form of current carried over speaker cables) into mechanical energy to drive the speaker cone, and the speaker cone itself serves as another transducer that converts mechanical energy into acoustical energy, or variations over time in air pressure. The sheet of paper represents yet another transducer that converts this acoustical energy back into mechanical energy. And this entire process began with electrons moving along a speaker cable!

Transducers can be decomposed into two broad categories: *sensors* and *actuators*. Actuators move things and do stuff: They like to convert energy from one form into mechanical energy. Examples of actuators include the voice coils in loudspeakers, electric motors, and hydraulic lifts. And just as the name indicates, sensors attempt to detect a physical quantity and convert it to a form that can be used by an electrical circuit or a computer; they come in hundreds of varieties and can be made to measure acceleration, tilt, rotation, gravity, radiation, light, color, weight, frequency, liquid flow rates, and even smell. The sensor we musicians are often most interested in is one specifically designed to measure variations in sound pressure (acoustical kinetic energy) over time and produce a corresponding output voltage: the microphone.

Introduction to Microphones

As you probably already know, many different types of microphones are available. These can be categorized according to the physics by which they work, their intended use, quality, size, cost, durability, features, popularity, and just about anything else. And microphones can cost as little as a dime or well over US$10,000. In this section, by way of introduction, we dissect some of the key terminology regarding microphones and then categorize them according to a very important feature: their directional sensitivity.

Terminology

Before we address specific microphone types and the basics of how to use them, it is essential to understand the terminology used in microphone parlance. When purchasing or selecting a microphone to use, one notices that they are typically tested and rated using several measures. These include the microphone's directional sensitivity (also called the pickup pattern or polar response), the physical process by which the microphone works, frequency response, transient response, self-noise, sensitivity, maximum sound pressure level, and impedance.

A microphone's *directional sensitivity* is one of the most important features to consider when selecting which microphone to use in a given situation. This refers to the device's relative ability to record sound sources in the plane around the front of the microphone—that is, how sensitive it is when recording sounds that are *on-axis* (directly in front of the microphone, defined to be 0°) relative to those that lie *off-axis*. Some microphones are highly directional, picking up sounds only in a narrow on-axis cone, while others can respond to sounds from any spherical direction more or less equally.

The directional sensitivity can be manipulated by the manufacturer to fit a desired specification in several ways. Variances in the microphone *capsule* that encases the internal part of the microphone can play a big role: Its size, shape, and whether or not it is ported to allow air from different directions to enter are all important factors. Likewise, the size, shape, makeup, and orientation of the microphone's internal *diaphragm*, which actually couples to and moves with incident sound waves, can affect the directional sensitivity, among other specifications as well.

Before sending a microphone to market, the manufacturer measures the directional sensitivity of the device and plots a *polar pattern* or *polar response chart*, like that shown in Figure 8.1. Note that the directional sensitivity of a microphone can also be frequency dependent: The precise polar pattern for one frequency band may be somewhat different from that for another.

Microphones are also rated according to whether or not they use *phantom power*, that is, whether an external power supply or battery must be used to make the microphone function properly. Whether a microphone requires phantom power is dictated by its physics of operation, as we will see momentarily.

Two of the most important ratings of a microphone—ratings that often differentiate the wheat from the chaff—are the *frequency response* and the *transient response*. (Loudspeakers are also rated in the same way.) The frequency response is a plot of the microphone's sensitivity (or gain, plotted in decibels on the y-axis) as a function of frequency (in Hz, plotted logarithmically on the x-axis). We are gen-

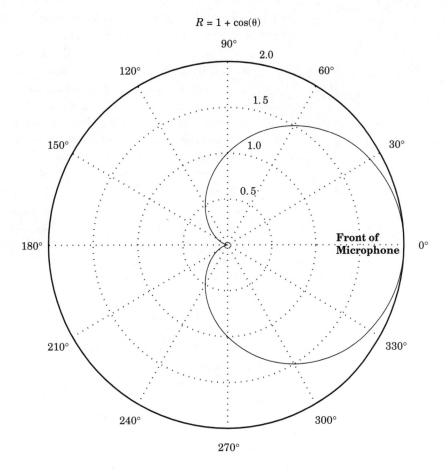

Figure 8.1
An example polar
pattern.

erally most interested in the microphone's on-axis frequency response, although some models give frequency responses for off-axis angles as well.

The transient response is a plot of the microphone's output response given an acoustic impulse, like a balloon pop. Ideally, given an extremely short sound—say, a 1 millisecond electronically generated noise burst recorded in an anechoic chamber—a microphone should produce a corresponding 1-millisecond output voltage spike, shutting off immediately following the end of the burst. However, because microphones rely on moving internal parts—in particular, a diaphragm suspended in air that vibrates in tandem with acoustic pressure waves—the output does not return to zero instantaneously. Instead, it oscillates a bit before resting and returning to its equilibrium (rest) position.

An ideal microphone does not color or alter the tone of the acoustic signal it is recording: It is far better to record a sound as faithfully as possible and "tune" it in software on the DAW. However, a microphone with a perfectly flat frequency response and impulse response is not physically possible. A real microphone's frequency response and transient response imbue different models with distinguishing tonal characteristics, some of them considered favorable for recording different sound sources—an often sought-after quality when one purchases a specific microphone.

Microphones also exhibit varying degrees of *self noise*, also called the *equivalent noise level*, which causes the device to produce small, noisy output voltages even in a perfectly quiet anechoic chamber. Self noise is caused by the microphone's internal electronics (largely the transistors or vacuum tubes inside) and to the random motion of ambient air that is inside the microphone capsule. Clearly, microphones exhibiting low self noise are preferred over those with higher self noise.

Another specification included with microphones is their *sensitivity*, which refers to the output voltage a microphone produces given a specific acoustic pressure. (Note that this is a different measure than a microphone's directional sensitivity.) Sensitivity ratings are usually given in millivolts (mV) of output voltage per Pascal (Pa) of air pressure at 1 kHz; alternatively, they may be given in the microphone's gain in decibels relative to 1 volt (V) per Pascal (Pa). For example, a sensitivity rating of 10 mV/Pa indicates that the microphone produces a 10-mV output voltage given a 1-Pa pressure wave vibrating at 1 kHz.

Microphones are also rated in terms of their sound-pressure level (SPL)-handling capabilities. Most microphones could not accurately record the sound emitted by a jet engine at close range; beyond a certain SPL threshold, the microphone's voltage output waveform distorts, eventually clipping because it cannot produce more than a given maximum output voltage. The maximum SPL a microphone can handle is usually given in terms of the corresponding *total harmonic distortion* (THD) produced given a specific SPL level. (THD is a quantitative measure of the amount of distortion of the overtone structure a system introduces to an input sound. More specifically, total harmonic distortion is the ratio of the total sum of the powers of all harmonics in a signal to the power of the fundamental frequency. An audio system—a microphone, in this case—with a very high THD rating, for example, would introduce harmonics into a pure sine wave.)

For example, suppose a microphone is given the following ratings:

134 dB SPL peak ($< 0.5\%$ THD)

This means that the microphone can record input sound pressure levels of up to 134 dB SPL while introducing no more than 0.5 percent total harmonic distortion. (This THD threshold of 0.5 percent is often used, as is 1 percent, as an approximate value for the threshold in which distortion becomes audible to most listeners.)

A related term that takes into account the microphone's self-noise and its SPL handling rating is the *dynamic range* of the microphone. The dynamic range is simply the difference in decibels between the maximum SPL rating of the microphone and its self-noise rating. Said another way, the dynamic range refers to the musically usable range of the microphone.

How can one ensure that the dynamic range of a microphone is being used to its full potential? Simply match the dynamic range of the source to be recorded to the dynamic range of the microphone as much as possible. In other words, if a quiet sound source is being recorded, consider placing the microphone closer and/or compressing the microphone's output before it enters the DAW to lower the noise floor. These actions serve to provide gain to the analog signal that the DAW's audio interface is asked to digitize, which of course minimizes the quantization error. (Recall from the first chapter that the lower the sampled amplitude, the higher to signal-to-error ratio.) In fact, inserting an analog compressor into the recording chain is typically done for just this reason.

Finally, as with other audio equipment, microphones are rated in terms of the *impedance* they exhibit. Impedance is simply the frequency-dependent resistance to the flow of electrons. Resistors "resist" this flow more or less equally for all frequencies, while things like capacitors and inductors favor some frequencies of electron flow more than others. Microphones can contain all three of these, and so they exhibit frequency-dependent resistance, which is "averaged" over the audio frequency range and measured in units of Ohms (Ω). Microphones can be categorized as low-impedance (low-Z), medium-impedance (medium-Z), or high-impedance (high-Z), as shown in Table 8.1.

TABLE 8-1

Microphone Impedance Ratings

Low-Z	Less than 600 Ω
Medium-Z	600–10,000 Ω
High-Z	Great than 10,000 Ω

In general, high-Z microphones are less expensive, but the length of the cable run used with them is more limited than with low-Z models. All things being equal, low-Z microphones are usually preferred for professional audio work; simply consult the microphone's manual for its specifications, or check the side of the capsule, which often lists the microphone's impedance. A rule of thumb is that high-Z microphones tend to use quarter-inch connectors (like the cheap models that the local electronics store sells), while low-Z microphones tend to use XLR connectors.

Directional Sensitivity Categories

Based on their pick-up sensitivity to the spatial location of sound sources, microphones can be classified into one of five broad categories: *Omnidirectional, cardioid, hypercardioid, shotgun,* and *figure-eight*. Most microphones are specifically designed to exhibit one of these polar patterns, although some (usually more expensive) models include a switch that allows the user to select the desired pattern for a specific application.

Omnidirectional ("omni") microphones exhibit a more-or-less circular polar pattern; that is, they are about equally sensitive to on-axis and off-axis sound sources. (See Figure 8.2.) Note that omnidirectional microphones become less...well, omnidirectional...when recording higher frequencies. This is similar to the head-shadowing effect mentioned earlier with the interaural level difference (ILD): Incident frequencies whose wavelengths are shorter relative to the microphone capsule cannot be diffracted around the microphone, thus limiting the diaphragm's ability to respond to them.

Cardioid microphones, as the name implies, feature a heart-shaped polar response pattern: They are most sensitive to on-axis and surrounding angles, less so towards the rear, and they are effectively insensitive to sound from directly behind (180° off-axis). They are used in situations in which it might be desirable to somewhat attenuate off-axis sounds from behind the microphone, like audience noise in a concert. They are also often used for recording vocal lines, because they tend to not exhibit the *proximity effect*, which causes other microphones to have an unnatural "boomy" bass response when placed too close to a source.

Rather than attenuating 180° off-axis sound sources, hypercardioid microphones exhibit nulls at ±45° off the rear axis (at 180° + 45° =

225° and 180° − 45° = 135°). Like cardioid microphones they are most responsive to on-axis and neighboring angles, but they also respond a bit to perfectly off-axis sound sources.

By including an acoustic waveguide tube in front of the capsule, shotgun microphones are designed to respond to on-axis sounds with as much off-axis rejection as possible. The tube, which is perforated along its length to allow off-axis sounds to enter and ideally destructively interfere with one another, can vary in length from short (around a foot) to long (close to three feet). A variation on the shotgun microphone, called the parabolic microphone, uses a parabolically shaped acoustic waveguide to direct on-axis sounds toward the diaphragm element. (This is especially great for eavesdropping!)

Finally, figure-eight microphones (also called "bidirectional" microphones) are designed to respond with identical sensitivity in the front-half plane and the rear-half plane, exhibiting sensitivity nulls at 90° and 270°. Figure-eight microphones are particularly susceptible to the proximity effect and rumble from wind, so their use is generally limited to studio-only recording applications. They are often useful in recording face-to-face interviews or pairs of instruments placed opposite one another (Figure 8.2).

Types of Microphones

Selecting the right microphone (or choosing one from among an appropriate set of choices) for a given recording situation is both art and science—often a matter of physics, acoustics, convenience, and personal preference. For any given scenario, a number of microphones might work well; again, patience coupled with trial and error are the best teachers.

Microphones can be classified into four types based on the physics of their operation: *dynamic*, *condenser*, *ribbon*, and *combination*. We discuss each of these types in turn first before moving on to microphone placement techniques.

Dynamic

Dynamic microphones work on the principle of electromagnetic induction. Simply stated, this means that a current running through a coil

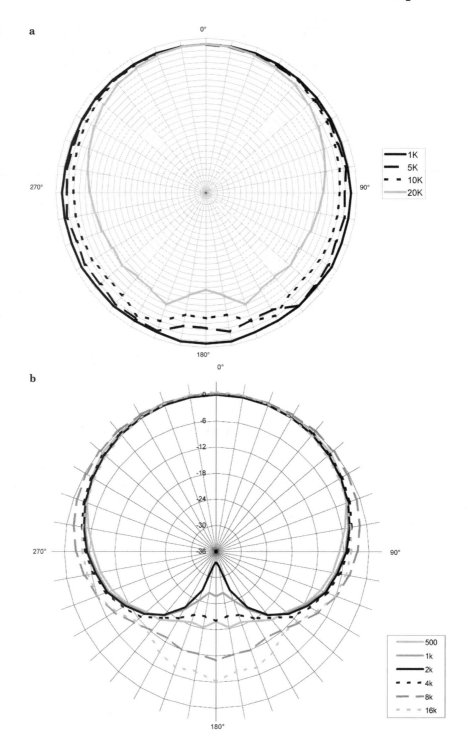

Figure 8.2

Four categories of polar response patterns: (c) hypercardioid (Earthworks SR-78); and (d) idealized figure-eight. Earthworks figures courtesy Earthworks Audio Products

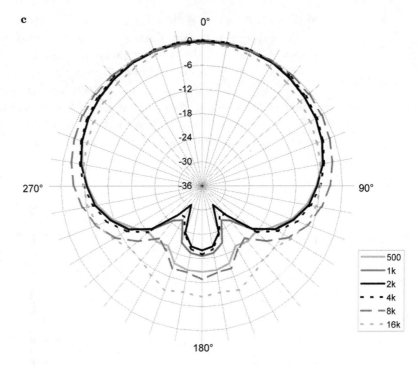

c

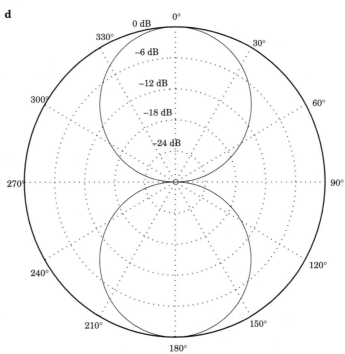

d

of wire induces a magnetic field, and vice versa. By attaching a magnet to a diaphragm and placing the magnet inside a coil of wire, the physical motion of the diaphragm (and hence the magnet to which it is attached) will induce current through the coil of wire. (See Figure 8.3.) This current runs through an electronic circuit inside the microphone, is signal-conditioned and amplified a bit, and converted to a voltage—the microphone's output. What causes the diaphragm to move? Incoming sound waves, of course!

Figure 8.3
The innards of a dynamic microphone. Mechanical motion of the diaphragm induces a current at the microphone's output.

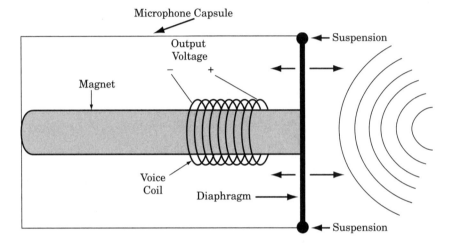

This is the same principle on which moving-coil loudspeakers work, but in reverse: By sending a current (from speaker wire) through a coil wrapped around a magnet that is attached to a diaphragm ("speaker cone"), the magnet—and hence the diaphragm—moves, creating sound waves.

Dynamic microphones are prized for their ruggedness, durability, and often lower cost. Additionally, they do not require phantom power to work, and they can be used with longer runs of cable. Although not usually as "accurate" as other microphones, they are often quite good at recording vocal lines, in field recording, and in live sound-reinforcement applications.

Condenser

Condenser microphones operate on the electrostatic principle of capacitance. Whenever two parallel plates face each other, an electric field can be generated between them by applying a voltage across the plates. For example, if one plate is connected to the (−) terminal ("ground") of a 9-V battery, and the other plate, placed a small distance away, is connected to the (+) terminal of the same battery (+9 V), then an electric potential energy field (i.e., voltage) has been placed across the plates, and charge builds up. That is, one plate becomes negatively charged, while the other will become positively charged.

The ratio of the amount of charge per unit volt that accrues is called the *capacitance* of the plates. Said another way,

$$C = \frac{Q}{V} \tag{1}$$

where C is capacitance (in units of farads), Q is charge (in units of Coulombs), and V is voltage (in units of volts). Basic electronics teaches another equation that describes capacitance:

$$C = \frac{k}{x} \tag{2}$$

Here, k is a constant; it is a number that depends on both the area of the capacitor's plates and the material between them, whether it be air, foam, or anything else. And here is where it gets useful for making a microphone: The variable x is the distance between the two plates. Thus, all things being equal, as the parallel plates move farther apart, the capacitance between them decreases.

A condenser microphone uses this equation in a clever way: By fixing one plate (called the backplate) to a rigid surface, we let the other plate (diaphragm) be mobile. The diaphragm, which moves in tandem with incident sound waves, causes the effective distance between the two plates to change. The time-varying capacitance that develops between the plates can be converted by a circuit inside the microphone into a time-varying voltage, which is the output of the microphone (Figure 8.4).

Recall that a voltage must be placed across the two plates in order for charge, and hence capacitance, to develop. This is where phantom power comes to the rescue. Condenser microphones require a bias voltage (typically +48 V) to work; this is the "V" in the equation $C = Q/V$. This voltage is also required to power an internal preamplifier inside the microphone's capsule to raise the voltage level a bit.

A special class of condenser microphones, called *electret microphones*, mitigates the need for such a high bias voltage by using a permanently charged backplate or diaphragm. A smaller battery is still usually required, however, to power the internal preamplifier.

Figure 8.4
The innards of a
condenser
microphone.

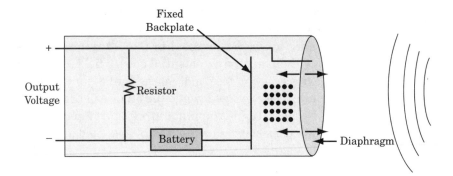

Condenser microphones are valued for the sonic accuracy, flatter frequency response, and good transient response. However, they are more fragile and generally more expensive than their dynamic microphone counterparts.

Ribbon

Ribbon microphones are based on yet another physical principle of operation: the piezoelectric effect (from the Greek word *piézo*, meaning pressure). Piezoelectric materials possess dipoles that are laid end to end (positive to negative), and any deformation of the material results in a net charge across the material. Conversely, any charge placed across the material results in a deformation of the material. The first property forms the basis of ribbon microphones; the second, electrostatic loudspeakers and ribbon tweeters.

A ribbon microphone uses a very thin film of piezoelectric material as its diaphragm. This yields two characteristic properties. First, such microphones are highly directional, responding primarily to on-axis sounds, because off-axis sound sources are less likely to cause the piezoelectric material to move much. Second, because the diaphragms are so thin (out of necessity, because if they are too thick, they will not deform very much), ribbon microphones tend to be quite fragile. They are also typically more expensive than dynamic and condenser microphones, although they generally possess the best high-frequency response.

Other Microphone Types

Other microphones are also available; they might use multiple dynamic or condenser "building blocks" encased within a single meta-

microphone, or they can couple existing microphone technologies with added functionality in some way. Examples of this type include stereo, dummy-head, SoundField, and digital microphones.

Stereo.   A stereo microphone combines two microphone elements into a single capsule that produces two output voltages: one for the left channel and one for the right channel. They offer ease of use in recording material in stereo, although they generally provide less flexibility in terms of placement than using separate microphones. Some newer stereo microphones, however, allow the relative positions of the individual internal microphones to be adjusted, thus lending greater flexibility to a number of recording scenarios.

Dummy-head.   A dummy-head microphone consists of microphone elements mounted inside the ears of a model of a human head. This allows the model's ear shape and separation to imbue a stereo recording with the natural timing, level, and tonal cues that a live listener might experience.

SoundField.   A SoundField microphone (http://www.soundfield.com) is a proprietary recording technology that facilitates ambisonic ("B-format") recordings. The microphone itself is highly robust and adaptable, as it can be used to record anything from mono point sources to three-dimensional surround audio information. It features four microphone elements mounted in a spherical array. And as ambisonic recording necessitates four channels (W, X, Y, and Z), a SoundField microphone produces four corresponding output channels, requiring a multichannel DAW (or a hardware multitrack recorder) to store its output. At its core, the SoundField microphone represents a three-dimension extension of the Blumlein stereo placement technique, which we talk about shortly, embodied in a single microphone.

Lavalier, headset, and wireless.   Lavalier, wireless, and headset microphones are tailored for specific sound reinforcement applications, although they can of course be used in studio recording as well in some situations. As the names suggest, a lavalier microphone is fastened near the neck of a speaker, typically to the lapel of a jacket. Headset microphones can facilitate recording the voice of instrumentalists while they are performing on their instruments. Finally, wireless microphones are convenient, eliminating the need for recording cables, but the recorded result is typically inferior to using a dedicated, wired microphone.

Digital.   Digital microphones simply include an A/D converter inside the capsule so that they produce a digital stream of numbers as their output rather than an analog voltage. The primary advantage that such microphones offer over analog-output microphones is increased noise immunity, particularly when their output is sent over longer runs of cable.

Finding transducers in weird places.   Some creative recording engineers have discovered transducers in strange places, using them to great benefit. Indeed, the best microphone for a particular situation might not be that expensive microphone in the closet. Some engineers, for example, like to record bass drums with a subwoofer. Recall that, to some extent, dynamic microphones and voice-coil loudspeakers are interchangeable. They are in fact the "inverse" of each other, but both rely on electromagnetic induction to operate. Incident low-frequency sound waves will cause a subwoofer to produce a current at its terminals that can be amplified and recorded. Others have explored using headphones as microphones and even produced homemade microphones. And it is somewhat in fashion now to use an analog telephone as a studio recording microphone, either by recording the direct output of the telephone line, or by disassembling the handset and soldering wires onto the internal microphone. Working in this way is done not to facilitate faithful reproduction of sound, but rather to impart tonal effects that might be otherwise somewhat difficult or more time consuming using traditional effects processing.

Primary Microphone Placement Techniques

Knowing where to place a microphone to obtain a quality recording requires knowledge of three things:

- The characteristics of the microphone
- The characteristics of the source to be recorded
- The characteristics of the room's acoustics

Coupling this knowledge with common sense and a delight in creative experimentation can take you a long way. We have spoken about microphone characteristics, including the basic physics of their operation and their directional sensitivity patterns, and we should all know something about room acoustics and related terminology by now (direct sound, early reflections, reverberation, parallel walls, and so on). Before undertaking a taxonomy of classic microphone placement techniques, however, we should address two important concepts relating to room acoustics and the characteristics of the source to be recorded (the second and third points of the list above).

Characteristics of the Source

Knowing the characteristics of the source to be recorded embodies three areas: the source's frequency range, its transient nature, and its three-dimensional radiation pattern. The first one is easy: The voice has a more limited frequency range, for example, than a symphony orchestra. Most recording engineers would thus not use the same microphone to record both.

The transient nature of the source can also indicate something about microphone placement. Human speech, for example, is full of quick transients and plosives. This would indicate that, to avoid the proximity effect, we might consider using a dynamic microphone with an omnidirectional polar pattern, being careful not to place it too close to the speaker. This also suggests that, owing to the "muddying" effect of early reflections and reverberant sound in rooms, we might want to place a microphone closer to its source to capture "cleaner" transients, or we might want to place it further away if we wish to soften them up a bit.

Finally, understanding something about the radiation pattern of the source to be recorded is critical. For example, short of creative license, one would probably not want to place a microphone directly behind the head of a vocalist to pick up a faithful recording—the sound is not radiating from there! The radiation patterns of acoustic instruments vary greatly; it is a good idea, until you are familiar with the best positions for placing microphones near instruments to achieve a faithful recording, to experiment with a microphone recording into a DAW to find the peaks and nulls of an instrument's spherical sound radiation.

Acoustical Considerations

It should be obvious by now that the closer a microphone is placed to a sound source, the greater the ratio of direct-to-indirect sound becomes. To obtain more "dry" signal and less "wet" signal (i.e., to minimize the recorded content of the room's acoustics), simply place the microphone close to the source. Conversely, to obtain more "wet" signal and less "dry" signal, move the microphone farther away. The direction the microphone is facing relative to the microphone's on-axis angle of course also affects the ratio of direct-to-indirect sound.

A term worth mentioning here is *critical distance*, defined to be the physical location in a room in which the intensity of the direct sound within the room is identical to the intensity of the corresponding indirect sound. In other words, the critical distance defines the location in which the direct sound and the sound that is reflected off surfaces in the room are equal in intensity. This location can be estimated by ear using a loud, noisy sound source placed at the front of the room, and backing away from it until no resulting loudness change is noticed. A much more accurate way, though, is to use a microphone and sound-analyzer plug-in for a DAW to digitally display the point at which movement of the microphone away from the sound source ceases to produce a corresponding drop in sound pressure level.

Ready to Record?

Microphone placement techniques can be classified in two ways: according to the physical location of the microphone(s) in an acoustic space, and according to the relative placement of two or more microphones with respect to each other. Microphone location options include *close*, *accent*, *distant*, and *ambient* placements. A separate issue particular to stereo and multichannel recording is the relative position of two or more microphones in the space.[1] Options include *spaced apart*, *semicoincident*, *coincident*, and *multi-microphone* techniques.

The discussion here is by no means exhaustive, but it should serve as a basis for further exploration and study. And the DAW provides a wonderful testbed for learning how to record. Multichannel, high-defi-

---

[1] With, of course, the exception of stereo microphones, in which two microphones are combined in one capsule, and SoundField microphones.

nition recording is easy with the DAW, and there is no need to keep purchasing tapes on which to record!

Just remember the golden rule of recording:

There are no rules. There are only good, better, and bad ideas.

Microphone Placement

As mentioned, microphone placement locations fall naturally into four categories: close, accent, distant, and ambient, in increasing order of distance from the sound source. When judiciously used with careful microphone selection and polar patterns, these placement techniques offer precise control of the ratio of direct to indirect sound.

Placing a microphone extremely close to a sound source (usually less than one foot or so away) is called close miking (also called "spot" miking). If the microphone is placed on a hard surface near the source, it is then called a boundary microphone. Boundary microphones, because they are placed as close as possible to a hard surface, are useful for minimizing the incident, indirect reflections.

Once a small microphone, like an electret element, is coupled directly to a vibrating body, like an acoustic guitar for instance, we call it a contact microphone. Contact microphones are used to record as much direct sound as possible while even further minimizing the incident, indirect sound.

Accent and distant microphone placement provide a successively greater level of indirect reflections (and hence room ambience) by moving the microphone back at least several feet from the sound source. Distant placement is generally used to capture the overall tonal quality of a sound source in its acoustic environment, possibly used in conjunction with an accent placement to pick up individual spot sources. These techniques are often combined when recording a symphony orchestra, for instance, in which distant microphones can record the overall ensemble, and accent microphones can record individual soloists. Note that we would not want to use a close placement on the soloists, because the natural ambience of the room would then not be recorded, making the soloists sound as if they were in a different environment.

Ambient microphone placement involves locating a microphone beyond the critical distance away from the sound source. In other words, ambient microphone placement is used to primarily record the ambience and reverberation within a room, not the direct sound.

A typical recording session may involve several microphones and multiple placements, even if the intended recording is eventually meant for stereo mixdown. Fortunately, using a DAW makes this process quite simple and the result very malleable: By recording the output of each microphone in real time to a multitrack editor, the dry-to-wet mix can later be adjusted in software by simply adjusting the relative levels of each track.

Any kind and number of microphones can be used in close, accent, distant, or ambient placement. But using two or more microphones raises several critical issues, to which our attention now turns in discussing three broad categories: spaced-apart pair placement techniques, coincident pair placement techniques, and multi-microphone placement techniques.

Spaced-Apart Pair Techniques

If we are trying to record a stereo image, one simple idea is to begin by placing two microphones some distance apart from each other. Our two ears, after all, are spaced apart, so this might be a good idea. In fact, spaced-apart placements are designed to pick up both the ILD and ITD cues that our ears receive. However, spacing multiple microphones apart when trying to record a stereo image has its drawbacks, most notably the likelihood of *phase cancellation*.

When two microphones are spaced apart, the acoustic pressure waves from any sound source not equidistant from both microphones will clearly reach each microphone at different times. Like we said, the goal here is to record ITD cues, something that would not be possible if both microphones were placed on top of each other (i.e., coincident). ITDs manifest themselves in a spaced-apart recording as phase differences between the two channels: If a sound wave reaches one microphone sooner than the other, the signal will be out of phase (not phase aligned with respect to the other channel). This is fine, and our ears indeed use these phase differences as a localization cue, but a big problem can occur if a particular frequency is around 180° out of phase in the two channels. In this case, the acoustic pressure rarefactions in one channel coincide with the compressions in the other, and vice versa, resulting in a virtual cancellation of that frequency. On stereo playback, that frequency becomes greatly or completely attenuated as a result. (See Figure 8.5.)

Figure 8.5
Phase
cancellations can
result from
spaced-apart
microphone
placement.

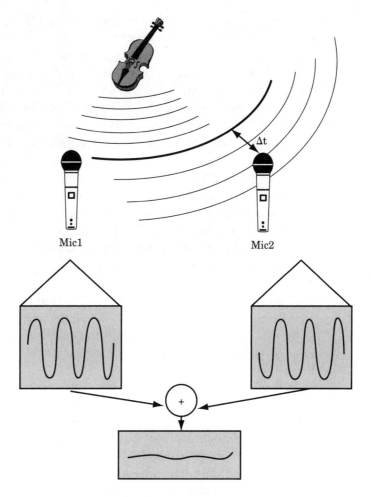

A common technique used to mitigate phase cancellations when using spaced-apart techniques, particularly when accent miking and multitracking sound sources that will be downmixed later to one channel, is the so-called *3-to-1 Rule*, which suggests that microphones be separated from each other at a distance at least three times greater than the microphone-to-source distance. This significantly reduces cross-talk (the leakage that "spills over" from one sound source to the other microphone) and thus audible comb filtering that would otherwise result when the outputs of each microphone are combined (i.e., mixed). For example, if you are spot-recording a cello and a clarinet, each with their own microphone placed 3 feet away from the respective instruments, then the two microphones should be placed at least 9 feet apart.

Spaced-apart techniques are also criticized for the ambiguity with which they encode the center of the sound stage, often manifest as a lack of "clarity" in the center image. Nevertheless, a well-constructed spaced-apart technique can result in remarkable stereo imaging, largely owing to its unique capture of both ITD and ILD information.

Spaced-apart placements fall into three broad categories: AB, ORTF, and binaural. Each involves a pair of microphones that are spaced anywhere from a few inches to several feet apart.

AB.   AB placement typically involves the use of two cardioid microphones either facing directly forward toward the stereo sound stage or perhaps angled slightly apart. Although particularly susceptible to phase cancellations, AB stereo recording can capture a broad stereo image in many live recording applications well and has its own loyal group of proponents.

ORTF.   The Office de Radio-Télévision Française (ORTF) has proposed a method for stereophonic recording; simply called the ORTF technique, it requires two cardioid microphones spaced 170 millimeters apart (about 11 inches) forming an angle of 110°. (See Figure 8.6.)

Several other ORTF-like offshoots were soon proposed by other European radio broadcasters. The NOS technique (after the Nederlandsche Omroep Stichting), requires two cardioid microphones spaced 300 millimeters (about 19 inches) apart, at an angle of 90°. In Italy, the Radio Televisione Italiana (RAI) offered a technique calling again for cardioid microphones but spaced apart 210 millimeters (about 14 inches) at an angle of 100º. In Germany, the Deutshes Institut für Normung (DIN) standard specifies two cardioids spaced 200 millimeters (about 13 inches) apart at an angle of 90º. A similar technique was proposed in the United States by RCA Engineer Harry F. Olson (of RCA Mark II Synthesizer fame), in which the cardioid microphones are spaced 200 millimeters (13 inches) apart at an angle of 135° Each of these results in a slightly different stereo image that can, of course, vary based on how far away the pair is located from the sound source.

Binaural.   Binaural microphone placements attempt to directly record the sound that would enter a listener's ear canals. Two techniques are possible: a dummy-head microphone can be used, or very small microphones can actually be placed directly inside a human

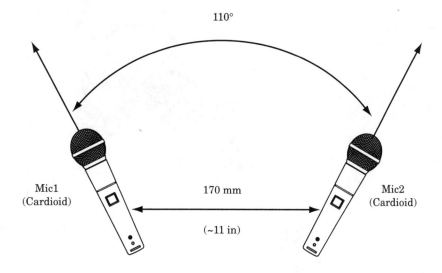

Figure 8.6
ORTF microphone placement.

subject's ear. Binaural recordings can work very well, being especially convincing over headphones, although the resulting image can degrade when such a recording is played back over loudspeakers. And using microphones placed inside the ears of human subjects of course requires that their heads stay precisely still, as any motion will either create rustling noise or skew the captured image.

Coincident-Pair Stereo Techniques

Coincident-pair stereo techniques attempt to mitigate the phase cancellation problem by capturing only ILD cues. Placing two microphones directly adjacent or on top of one another allows the resulting recording to encode ILD cues only; no ITD cues are possible, because the microphones are coincident. A potential disadvantage, however, is that the stereo image may suffer somewhat owing to the lack of ITD information.

XY.   One of the more commonly encountered stereo recording techniques is the XY-pair configuration, which calls for coincident cardioids angled at 90° (Figure 8.7). The result looks something like the letter "X." Following on the idea of the single stereo microphone, some newer microphones include a built-in XY-pair arrangement incorporated into one unit.

Figure 8.7
XY microphone
placement.

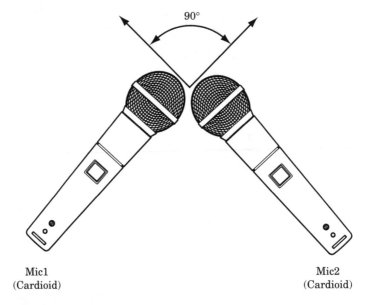

Mic1
(Cardioid)

Mic2
(Cardioid)

Blumlein   In 1933, the British engineer Alan Dower Blumlein (often called the "inventor of stereo") obtained a patent (one of over a hundred he obtained during his lifetime) for stereo recording by using figure-eight microphones angled at 90° with respect to each other and placed one on top of the other. (See Figure 8.8.) The resulting configuration, called the Blumlein technique (big surprise!), forces the on-axis pickup of one microphone to coincide with the null of the other, in both the front and rear, resulting in a very stable image. This almost, but not quite, amounts to something like "omnidirectional stereo," something few other placement techniques can claim.

M/S.   In the same 1933 patent, Blumlein also mentioned another robust stereo recording configuration. Placing a single microphone (of any polar pattern) facing forward with a coincident figure-eight microphone angled at 90° facing left results in the Mid-Side (M/S) stereo-pair configuration. Consider the effect of this arrangement: one microphone is devoted solely to recording the front-facing, center image, and another captures only the two sides. Thus, the resulting output of the microphone pair does not correspond to the typical left and right channels we have come to expect.

Rather, the two output channels must be decoded, for example using a DAW plug-in, to achieve a left-to-right stereo image. An M/S decoder works on the premise that the two lobes picked up by the fig-

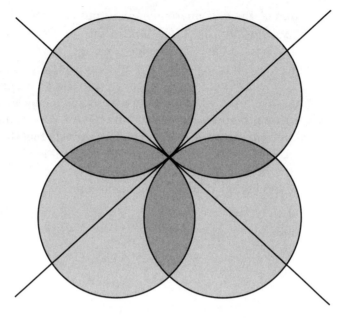

ure-eight microphone are precisely 180º out of phase. This makes sense, because each lobe of the figure-eight microphone "sees" the inverse of what the other "sees." In other words, a rarefaction in air pressure on one lobe necessitates a compression in air pressure on the other side, and vice versa.[2] By using this feature of the figure-eight microphone, an M/S decoder need only do two things:

- By adding the output of the side-facing figure-eight microphone output of the front-facing microphone, a virtual left-facing image is created.
- By adding the *phase-inverted* output of the side-facing figure-eight microphone to the front-facing microphone, a virtual right-facing image is created.

In simpler language, Left = $M + S$, and Right = $M - S$. (Note that the phase-inversion in creating the right image is accomplished simply with subtraction.) The cool thing about this is that the mid and side portions can be arithmetically weighted in different ways on a DAW, creating an infinite variety of stereo imaging possibilities. For

---

[2] Still not convinced? Remember that a typical figure-eight microphone only contains one diaphragm. This, whether it is moving inward or outward, depends on the perspective from which it is viewed (from the front lobe or the rear lobe).

example, the side microphone's output channel might be attenuated a bit in the M/S-decoding plug-in on our DAW, resulting in a more front-facing image. Conversely, we might want to widen the stereo image by attenuating the mid microphone's output channel.

Another advantage inherent in the M/S technique is its mono compatibility, even after decoding, which is something few other stereo recording techniques can claim (Figure 8.9). Consider the effect of adding the decoded left and right channels together: We get $(M + S) + (M - S) = 2M$. The effect of the addition is to double the amplitude of the mid channel and eliminate the side channel, which yields a phase-coherent center-facing mono channel.

Figure 8.9
M/S microphone placement.

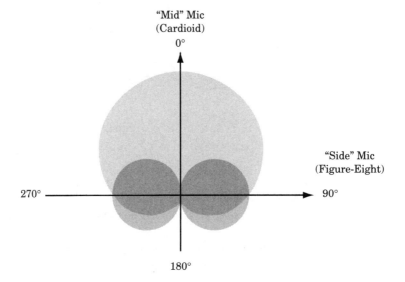

Multiple Microphone and Surround Recording Techniques

Using more than two microphones to record sound is often born of an attempt to combine the advantages offered by spaced and coincident arrangements, such as capturing a variety of ILD and ITD cues as well as a strong center image, so that the results can later be mixed and tweaked in software. And short of using a SoundField microphone or binaural techniques, the use of three or more microphones is required for accurately capturing discrete channels of surround audio.

Multi-microphone recording techniques are especially facilitated by direct-to-disk recording to a DAW. Recordings from any number of microphones can be captured to disk in real time, provided the DAW's hard disk is fast enough and the audio interface contains ample input channels. The only real potential drawbacks to multi-microphone techniques are the added expense and time involved. We briefly survey a few of the many possible multi-microphone arrangements.

One of the most famous multi-microphone recording arrangement is the Decca Tree. Invented in 1954 by recording engineers working for the Decca record label, the technique is primarily used to capture classical music performances in large venues, and it has remained largely unchanged for several decades. Three omnidirectional microphones are placed in a more-or-less equilateral-triangular arrangement, about 70 to 100 cm (about 45 to 65 inches) apart from each other, with one in front of the other two and closer to the sound source. Because the front microphone is centrally located and placed closer to the sound source, the precedence effect helps ensure a focused center image, something spaced-apart techniques alone generally do not offer. In a sense, the Decca Tree represents the best of both the spaced-apart and coincident worlds.

The Decca Tree itself is arranged on a large microphone array beam and placed typically 10 to 12 feet above the level of the stage. It is also often augmented with additional side-flanking microphones placed toward the edge of the stage, and also typically with accent microphones as well. The classic Decca Tree configuration has recently been adapted for surround recording: Replacing or augmenting the omnidirectional microphones with an M/S pair or a SoundField microphone, for example, can easily facilitate large-scale surround audio recording.

Recently, the traditional M/S pair has been adapted for recording surround audio with the so-called "Double-M/S" technique. By simply adding a rear-facing mid microphone, 5.1-channel recording becomes possible using only three microphones. Assuming M_{front} represents the front-facing omnidirectional microphone, M_{rear} represents the rear-facing omnidirectional microphone, and S represents the side-facing figure-eight microphone, we can decode a 5.1-channel recording thus:

$$L = M_{front} + S$$
$$C = M_{front}$$
$$R = M_{front} - S$$

$$L_s = M_{rear} - S$$
$$R_s = M_{rear} - S$$

The LFE channel can simply be taken as a low-pass-filtered copy of some combination of the resulting channels. Furthermore, as with traditional M/S recording, the "spaciousness" of the resulting image can be adjusted by tweaking the relative levels of M_{front}, M_{rear}, and S in the decoding process.

Many recording engineers are experimenting with other multi-microphone surround recording arrays. These include arrangements such as Thiele's Optimized Cardioid Triangle (OCT) Surround, IRT Cross, Hamasaki Square, Ideale Nieren-Anordnung[3] 5 (INA 5), Fukada Tree, and spherical-microphone systems.

Recording to a DAW

Using a DAW to record audio offers many advantages over traditional tape-based systems. First, there is no need to buy tape for a dedicated digital or analog tape recorder. Second, hard disks are more stable over a longer time than is tape. Third, a sound file can be duplicated much faster than a tape can typically be copied. And thanks to things like DVDs, laptops, and tiny flash-based drives, recordings and mixes can easily be transported from one place to another.

Another recording area in which DAWs are particularly adept is in overdubbing, punching, and inserting. Using a RAM-based loop recording tool, like MOTU's Polar, enables thick textures to be continually overdubbed. And because hard-disk recording encodes bits on a circular disk inside a computer rather than magnetic patterns on a linear tape, random-access editing is possible. In other words, the rewind and fast-forward operations happen instantaneously.

This chapter concludes with a discussion of DAW-specific recording issues, including latency considerations, preamplifiers, recording in the project studio, and the advantages of the laptop-based DAW in recording applications.

---

[3] "Ideal Cardioid Arrangement."

Recording Latency Considerations

Before recording with a DAW, it is important to consider latency issues, especially if a studio musician will be monitoring the mix over headphones as it is recorded. All digital audio interfaces introduce a small delay, or latency, between the time that a microphone picks up a sound and the time that the audio is quantized and recorded to the hard disk. This latency, typically on the order of milliseconds, is due primarily to the sample-and-hold circuits that A/D converters rely upon to work. (By analogy, think of a camera when it is taking a snapshot: It takes a small amount of time for the shutter to open and close back again.) The latency introduced by a DAW audio interface is a published specification of each model and should be checked before purchasing one.

Again, this latency is normally not a big issue unless the musician is monitoring the mix over headphones. And the latency introduced by the audio interface can potentially be multiplied considerably when software plug-ins are invoked, or if the DAW's hard disk is not fast enough to keep up with the number of channels being recorded in real time. Three simple ways to effectively remedy this situation are as follows:

- Most DAW audio interfaces include a dedicated analog headphone output jack. This is typically wired internally directly to a pair of the audio output channels, designed to route the input signal to the interface directly back out with zero latency. Using this headphone jack as the monitoring source for a headphone mix rather than one of the other output channels can eliminate any perceived delay between the performer's physical actions and the sounding result over headphones.
- Divide up the recording tasks by using another dedicated DAW exclusively for real-time plug-in processing of a recorded input. This can minimize latency, because the tasks of effects signal processing and hard-disk recording are decoupled, diminishing the overall processor burden.
- Record fewer channels at a time, or use fewer plug-ins at a time.

Again, if a computer forms the centerpiece of the studio, it makes sense to get the best model you can afford.

Preamps

Microphones require a preamplifier ("preamp") in order to work, because they produce such low output voltages owing to the physics of their operation. Microphone level ("mic level") is typically on the order of 2 to 20 mV, whereas the so-called "line level" that most DAW audio interfaces expect at their inputs is closer to 1 V, a factor of 50 to 500 times higher. As such, microphones must be used with a preamplifier, which can be constructed using analog vacuum tube circuitry (prized for its often favorable coloration and subtle analog noise) or discrete transistors (prized for their transparency, lack of coloration, more linear operation, and lower noise floor). Note that some audio interfaces for the DAW include a built-in preamplifier, so shop around when comparing models. Alternatively, standalone universal serial bus (USB) and FireWire (IEEE-1394) preamplifiers are also available; these allow their output to record directly into a DAW.

Again, augmenting a DAW with a small external mixing console—whether analog or digital—may help here, because most mixers include internal preamplifiers. Used in this way, a microphone is plugged directly into the mixer, and the mixer's output is fed into the DAW audio interface.

Recording in the Project Studio

Home recording presents many challenges; chief among them is the difficulty in finding a quiet location in which to record. And finding a suitable room that lacks slap-back echo and tonal coloration is typically even more difficult. Short of converting a bedroom or the garage into a dedicated, acoustically treated room for recording, two other cost-effective options are available.

First, many project studio-based musicians simply record directly into their DAW as much as possible without using separate microphones. This is great for electronic and amplified instruments that can send an analog or digital audio signal directly into the DAW's audio interface, like electric guitars, synthesizers, drum machines, samplers, and so on. Then, for any acoustical recording, a recording session is booked on an hourly basis in a local recording studio. Working in this way can minimize the amount of costly in-studio time required, because most of the recording and mixing can occur back in the project studio. It also affords the opportunity to record in a better acoustical space and typically with the assistance of a more experienced recording engineer (not to mention an assortment of microphones better than most of us can afford).

Another option aside from converting a room at home is to create a makeshift iso-booth in a closet, preferably one located in the interior of the house (and thus farther away from outside noise). The clothes already in the closet can help to absorb reflections for tracking in which a minimum of acoustical coloration is desired, enabling ambience to be added artificially on the DAW using a plug-in.

The Laptop Studio Revisited

The laptop-based DAW deserves special mention here, for augmenting a laptop with a multichannel audio interface and a multitrack editor can easily turn it into a portable, robust recording rig. The same laptop used in the studio can easily be transported, making the laptop DAW the ideal field-recording device. Future fuel-cell laptop batteries that promise days of continuous use can even make the ultimate portable hard-disk recorder more appealing.

For Further Study

Alexander R. *The Inventor of Stereo: The Life and Works of Alan Dower Blumlein*. New York: Focal Press; 2000.

Alldrin L, et al. *The Home Studio Guide to Microphones*. MixBooks; 1998.

Berke T, et al. 1993. "DAT-Heads Frequently Asked Questions: Microphone Edition." http://www.harmony-central.com/Other/mic-faq.txt.

Burns R. *The Life and Times of A. D. Blumlein*. IEE Publishing; 1999.

Huber DM, Runstein R. *Modern Recording Techniques*, 5th ed. Focal Press; 2001.

http://www.sfu.ca/sca/Manuals/ZAAPf/

http://www.mediacollege.com/

Streicher R. "The Decca Tree." *Mix Magazine* 2000; September.

"M-S Stereo: A Powerful Technique for Working in Stereo" published in the *Journal of the Audio Engineering Society*, 1982;30:10, 707–718.

Theile G. Multichannel natural music recording based on psychoacoustic principles. AES Convention, 2000, Preprint 5156.

Exercises and Classroom Discussion

1. Record as many sounds as you can possibly make with a piece of paper into a DAW.
2. Record a short musical passage played by an instrumentalist on stage using three different microphone placement techniques. Compare and contrast the results when listening (a) over loudspeakers and (b) over headphones.
3. Take apart a cheap microphone and put it back together.
4. Locate the specifications and read reviews of five different microphones in the same price range. Compare their features and users' comments.
5. Using anything except a traditional microphone, record as many sounds as you can possibly make with a drum into a DAW.
6. Record five sounds (human, mechanical, environmental, or otherwise) whose source of production will be unclear or ambiguous to a first-time listener.
7. Record impulse responses (to as many channels as you can) from three different locations within the same acoustic space. Compare the resulting responses, and see if a listener can tell the locations where each was recorded.
8. What was the earliest recording ever made?
9. Obtain a copy of the earliest recording you can find and play for the class.
10. Arrange a performance of Steve Reich's *Pendulum Music* (1968), and record the performance directly to a DAW.
11. What advantage and disadvantages do digital microphones offer? What about digital power amplifiers, for that matter?
12. Play a game of *Name That Microphone*! Pair up with a friend, and record the same sound source using five different microphones. See if, with practice, you can blindly discern which microphone recorded the sound on playback.

Expanding
the Palette

As stated before, one of the primary advantages of the DAW lies in the fact that a general-purpose, expandable computer lies at its core. Owing to this feature, it can take advantage of many other software programs that have been developed for other applications besides digital audio. By integrating the musical instrument digital interface (MIDI) protocol, digital video, software synthesis, computer-assisted composition, and an unbounded number of other possibilities into a central audio workstation, the traditional role of the DAW can be easily expanded to assist in an infinite variety of other creative tasks.

This chapter is divided into six main sections: synchronization, MIDI, software sound synthesis, computer-assisted composition, and digital video. We examine some of the principal components and concepts involved with each of these areas, concluding with remarks about the potential musical roles that the DAW might play in the future.

Synchronization

Because the DAW encapsulates so many of the functions normally provided by multiple machines in a traditional studio, the need to synchronize a plethora of digital audio (and video) equipment is drastically minimized. In a traditional, console-based studio, for example, all digital equipment must be synchronized to a common clock: the effects processors, modular digital multitracks, DAT/CD/DVD recorders, video recording/playback decks, and of course the mixing console, if it too is digital. A well-equipped DAW can perform all of these tasks internally as well, and so the only synchronization that may be used is taken care of internally, for example, by time-aligning a video clip for which one is composing the soundtrack in a multitrack mixer.

However, many DAW-based studios, particularly professional or postproduction studios, must interface the central DAW to other studio equipment, either for legacy compatibility issues (e.g., ripping a client's ADAT tape into Pro Tools), or for augmenting the central computer's functionality in some way, for instance with a hardware mixing controller. In such instances, a stable, common timing reference must be used so that all digital equipment agrees not only on the sampling rate, but—just as important—on the exact moment in time that each clock pulse occurs. Doing so allows bit-for-bit perfect recordings, copies, and data transfer among all devices in the studio chain.

Synchronization signals come in two basic flavors: *word clock* and *timecode*. Word clock is simply a "dumb" timing pulse: It's the metronome of the digital audio orchestra that tells all digital equipment the precise moments in time that digital words can be sent or recorded. The faster the word clock, the higher the sampling rate, and vice versa. Word clock is commonly used to tell all digital audio equipment the sampling rate the studio currently uses. Timecode, on the other hand, is more of a conductor than a metronome. It not only indicates the timing of the digital audio words but also the downbeats and measures, so to speak. Both word clock and timecode are commonly used in digital studios, and we talk about each in turn.

Word Clock

If all digital equipment in a studio only agreed on a common sampling rate, but not the exact time of each sample tick, problems could still occur. This scenario would be analogous to asking two musicians to play together by giving each an ideal metronome set to the same tempo marking, but not starting them at the same time. We could manually set a modular digital multitrack recorder and our DAW's audio interface to both run at a sampling rate of 96 kHz, for example, but without agreement on the exact timing of the words, then clicks and pops would most likely be introduced when recording from one to the other. This can occur because the transition times of the digital audio words are literally out of sync with each other: The transition times between words always occur at different times with respect to each other, which can cause audible problems with D/A and A/D converters (Figure 9.1).

Most professional digital audio devices are capable of generating or receiving word clock synchronization signals. By necessity, only one device can generate the word clock signal, and it is designated the *word clock master* or *house clock*. Recipients of the house clock are set to a *slave clock*. In project studios, the DAW's audio interface is most often designated as the master, and all other devices are chained together in slave clock mode. In larger studios, a standalone word clock distribution amplifier may be required to send the house clock to many slave clock devices.

In general, the ideal word clock synchronization signal is a simple 5-V peak-to-peak square wave, the same pulse signal used for decades in synchronizing TTL logic circuits. It is usually carried over a BNC

Figure 9.1
Synchronization
helps ensure that
each connected
digital audio
device agrees on
sample ticks.

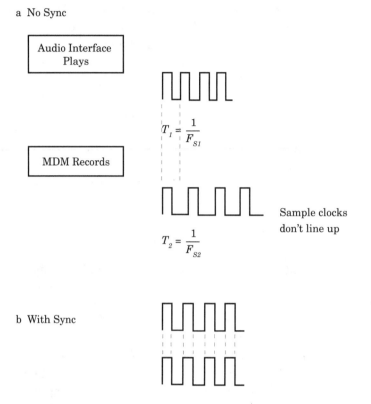

a No Sync

Audio Interface
Plays

$$T_1 = \frac{1}{F_{S1}}$$

MDM Records

$$T_2 = \frac{1}{F_{S2}}$$

Sample clocks
don't line up

b With Sync

cable. Nevertheless, errors may occur in transmitting or receiving the word clock signal; these can be caused by anything from impedance mismatches, abnormally long cable runs, or using bad cables. Timing errors can also occur: In particular, *jitter* results when the timing pulse is late or early with respect to its nominal timing location. The amount of jitter exhibited by any digital audio device, although often on the order of only a few nanoseconds but potentially a great deal higher for "prosumer" audio equipment, must be considered. Obviously, a lower jitter specification is desirable.

Because many different electronic techniques are available for generating word clock signals, some companies have developed proprietary implementations of word clock formats. The industry-accepted formats are the general-purpose "Word Clock" standard, which uses 5-V square waves as mentioned, and the AES11 specification, which allows word-clock synchronization information to be sent over AES digital audio connections. Other proprietary formats include Lucid's UltraClock and Digidesign's SuperClock.

Timecode

Using timecode enables digital devices to not only communicate synchronously, but, by keeping track of absolute time, it provides a control mechanism for instant and random-access playback, recording, and editing. Because timecode can assign a unique identifying index to each audio and video frame, whether the information is stored on tape, disc, or inside a computer, it is most often used in studios that work with audio and video material simultaneously. Just as word clock is used to synchronize the timing of digital audio words and sampling rates of multiple digital audio devices, timecode is used to synchronize audio and visual data streams both with respect to each other inside the DAW and across multiple computers, video decks, and recording equipment. Using timecode is thus essential in audio-video postproduction studios.

Several timecode systems exist, including ADAT Time Code and MIDI Time Code. But by far the most commonly used timecode system today was developed in 1967 by the Society of Motion Picture and Television Engineers (http://www.smpte.org).

SMPTE.   Before SMPTE timecode, the primary technique used for time-aligning multiple audio/video film or tape reels was to mechanically synchronize the playback rates of sprocket holes (in the case of film) or to encode a fixed-frequency tone (such as 60 Hz) onto video tape. In the latter case, a frequency-detection circuit on the playback device could analyze the encoded tone and make adjustments to the speed of the tape as necessary.

SMPTE Time Code assigns a unique identifier to each audio-visual frame by encoding absolute time information, that is, the absolute time relative to the very first frame. Specifically, SMPTE Time Code calls for time to be encoded in *hours:minutes:seconds:frames* format. In other words, the exact number of hours, minutes, seconds, and frames that have passed since the very first video frame is recorded onto each successive frame.

What are frames, anyway? Borrowing from the film industry, a frame, as we've used the term here, is simply a single still image. When we see a critical minimum number of frames played back in sequence per second, our brains can be led to interpret smooth motion; in other words, we can interpolate the in-between frame motions, and we want them to continue in the same direction. (Gestalt psychologists call this the principle of good continuation.)

(As an aside, recall that we have previously used the term *frame* to denote a collection of adjacent digital audio samples. For example, a single audio frame in a stereo sound file would comprise the audio words in each of the two channels.)

Getting back to video and film, picture frames of 24 to 30 frames per second (fps) are quite common and are adequate to impart a sense of continuous motion on playback. SMPTE allows frame rates of 24, 25, 29.97, and 30 fps, which are most often found in commercial film, European television (phase alternating line, or PAL), American television (National Television System Committee, or NTSC), and older NTSC equipment, respectively.

Because SMPTE Time Code has most often been recorded onto film and video tape, several techniques for recording the timecode onto these media have been developed. These include linear timecode (LTC) and vertical interval timecode (VITC). LTC is a digital timecode signal that is modulated onto a carrier in the audio range so that it can be stored on the analog soundtrack of the storage medium. VITC, on the other hand, encodes digital timecode information onto the vertical-blanking interval of analog video signals. (The vertical-blanking interval is the short time in which the electron gun in a traditional CRT television stops shooting electrons at the screen so that it can reposition itself at the top of the screen to "draw" a new video frame. This is the space in which closed-captioning information, for example, is stored in analog television signals.)

With the slow but steady decline of tape-based recording media in the audio industry (and now the video industry as well), however, LTC and VITC are becoming something of a vestige. SMPTE Time Code can easily be recorded onto digital recording media like DVDs. And the proliferation of newer, robust audio and multimedia encoding formats, such as the next-generation MPEG standards, allow timecode information to be stored in the digital bitstream itself, much as audio encoding systems like Dolby Digital and DTS already do.

On a final note, with the variety of word clock and timecode formats in use, some studios use dedicated hardware to convert among the various formats as necessary. Such devices, called synchronization converters, are often found in commercial studios and in post-production environments that routinely work with a variety of media and audio/visual formats.

Robust synchronization of the myriad digital devices present in many studios is indeed paramount to successful integration of the entire studio into a single instrument. But whether or not the typical project DAW-based studio requires word clock- or timecode-based synchronization, at some point, most users will encounter the now-ubiquitous world of MIDI.

MIDI

In many ways, the Musical Instrument Digital Interface (MIDI) marks a turning point in the history of electronic music. Arguably one of the first "standards" in the field, MIDI marked the demise of proprietary, nonstandard studio control paradigms and the beginning of cross-platform, device-independent control of electronic musical instruments. It also delineates the genesis of a ubiquitous, commercially accepted representation standard for musical instruments, remarkable especially because the standard has now survived over two decades.

Before MIDI, synthesizers and electronic music devices routinely communicated with each other using *control voltages*, a now-classic technique widely used by Robert Moog in his analog synthesizers. In fact, the term "voltage-control synthesizer" was applied to analog synthesizers of the 1960s and 1970s, allowing them, unlike their predecessors, to be predictably and reliably played remotely using a control voltage input.

For example, if a synthesizer player wanted to control other synthesizers in tandem (especially useful, because virtually all early synthesizers were monophonic; that is, they could only play one note at a time), the control voltage output of the *master* synthesizer would be "patched" into the control voltage input of the *slave* synthesizer. A control voltage of 0 V could instruct the slave synthesizer to play its low-C note, while a control voltage of 5 V could instruct it to play its high-C note five octaves higher.

The control voltage idea was great for many reasons: It was a control mechanism that was easy to implement using well-understood analog circuits, and, perhaps more important, it was an analog signal, which allowed arbitrary and expressive pitch bending and glissando. Indeed, one of the classic sounds routinely associated with the voltage-control synthesizer is the characteristic note-to-note portamento that resulted from analog voltage control.

Since the introduction and widespread acceptance of MIDI as a standard, many advances in digital audio technology have occurred that seem make the MIDI standard doomed to obsolescence. Regardless, it does indeed serve its designed purpose and then some, and the standard, despite its limitations, seems here to stay for at least a good while longer.

This section begins with an overview of the MIDI specification, including just a few of the more important details. We then move to a discussion of extensions to the specification, followed by the idea of the MIDI controller and its place both in the DAW-based studio as well as DAW-based live performance. The section concludes by examining some of the post-MIDI control specifications that have been proposed.

An Overview of the MIDI Specification

MIDI is an asynchronous, serial communication standard and accompanying hardware interface specification that allows electronic music gadgets to "speak" with each other. The standard is officially maintained by the MIDI Manufacturers Association of America (http://www.midi.org), composed of a loose affiliation of manufacturers of electronic musical instruments and DAW software. The fact that it is serial means that bits of information can be sent in sequence over one wire, as opposed to parallel communications standards, which require multiple (parallel) wires to transmit data. That it is asynchronous means that master and slave devices need not be time-synchronized with a common clock pulse, as is the case with digital audio devices requiring word clock synchronization. Communication can only occur in one direction over a single MIDI cable (up to 50 feet maximum in length, terminated in a 5-pin DIN connector); that is, separate input and output ports are required for bidirectional communication. In fact, most devices include three MIDI ports:

- **MIDI In Port:** Receives MIDI messages
- **MIDI Out Port:** Sends MIDI messages
- **MIDI "Thru" Port:** Duplicates the messages received at the MIDI In port and sends them back out the device

The "Thru" Port allows one master device to control multiple slave devices in a daisy chain formation by echoing MIDI messages to other devices in the chain.

MIDI, as you probably already know, is ubiquitous: The standard is implemented on most of the hardware devices and software programs found in modern-day studios of all levels, from low-end home studios to high-end production facilities. Using this standard, a keyboard controller can send information about how it is being played—what keys are pressed, when they are pressed, how hard, and so on—to a DAW; likewise, the DAW can play the keyboard. In fact, any MIDI-compliant devices can communicate: computers, instruments, modular digital multitrack recorders, or anything else on which the standard has been implemented. And although some analog synthesizers, like those made by Don Buchla, did include sequencers capable of storing, recalling, and performing sequences of notes, the advent of the MIDI specification engendered a new era of robust digital sequencers, both in hardware and software. The legacy of these devices and programs is today manifest in the powerful multitrack editor-sequencers available for the modern DAW.

But sequencers are not the only programs to benefit from the advent of MIDI. A great many other technologies (some of them not even musical in nature, but we'll get to that in a minute) rely heavily on MIDI to operate. One of the most powerful of these is desktop musical notation. By supporting the MIDI specification, notation software can record the live input of a performer as notes on the page or play the entered notes back to a musical instrument.

The MIDI standard calls for MIDI-compliant devices to communicate at a rate of 31,250 bits of information per second (baud). That's awfully slow compared to the bandwidth allotted to newer serial standards like universal serial bus (USB) and FireWire (IEEE-1394). But regardless, it seems remarkably resilient to carry the representational musical information for which it was originally designed.

Tracks and Channels

Sometimes confusing to beginners is the distinction between *tracks* and *channels*, analogous in many ways to their distinction in multitrack mixing software. The MIDI specification allows a single cable to address up to sixteen discrete channels of information. In a studio with two synthesizers, for example, we might program one to respond only to Channel #1 and the other to respond only to Channel #2, thus enabling the DAW to directly control each machine. Or, as is more commonly the case, one synthesizer may respond to all sixteen channels, with a different timbre ("program") on each channel.

However, *tracks* are a software-specific concept found in multitrack editor-sequencers—and not addressed in the MIDI specification per se—that can make it easier for us to create and modify MIDI information. The information contained in one discrete MIDI channel may be comprised of several tracks. And just as individual audio tracks must be assigned an output bus or channel in multitrack software, so too must MIDI tracks be assigned to a specific MIDI channel.

It might be helpful to think of the distinction in terms of a symphony orchestra. Each instrument family could be represented by a channel, while the individual instruments within the family may be represented as tracks. For example, Channel #1 may represent the violins, while the first and second violins might each be recorded on a separate track.

MIDI Messages

Although most of the nitty-gritty MIDI-specific implementation details are taken care of within commercially available hardware and software, it helps to get a sense of some of the lower-level operations involved in MIDI communication. At the root of all MIDI control and communication is the MIDI *message*, which is simply a two- or three-byte string of data. Whether the message is comprised of two or three bytes depends on the kind of message. Each MIDI message contains a *status byte* followed by one or two *data bytes*. Status bytes essentially tell the recipient, "Hey, I'm about to tell you some specific information about something in particular." Data bytes, on the other hand, say, "OK, and now here is the stuff I am going to tell you."

Recall that a byte represents a collection of eight bits, but the MIDI specification requires each byte to have its own unique *start bit* to identify the byte as a status byte or a data byte. A start bit of 1 indicates a status byte, whereas a start bit of 0 indicates a data byte. Because a start bit is required for each bit, only seven bits remain in each byte to store information. This is why MIDI information is limited to numbers in the range of 0 to 127, a total of $2^7 = 128$ quantization levels, unless the most- and least-significant bytes of 14-bit numbers are dissected and sent separately.

Status bytes include control information like "Note On" and "Note Off," meaning "I'm about to tell you which note to turn on" and "I'm about to tell you which note to turn off." The data byte, which immediately follows the status byte, indicates specific information regarding the status byte just sent, for example, the specific note to turn on and with what amplitude.

Note that MIDI devices are not required to respond to all possible MIDI messages to be considered MIDI compliant. The exact features of the standard that a specific device implements (that is, a listing of the messages to which it does in fact respond), are typically provided by the device's manufacturer in the form of a *MIDI Implementation Chart*.

Three basic types of MIDI messages exist: *Channel voice messages*, *channel mode nessages*, and *system messages*. We now discuss each of these in turn.

Channel voice messages. Channel voice messages, perhaps the most commonly encountered kind, provide control information intended for a specific MIDI channel. Seven different channel voice messages are available, their names revealing something of the keyboard-centric bias of the MIDI specification: *Note on*, *note off*, *polyphonic key pressure*, *channel pressure* (also called *aftertouch*), *program change*, *control change*, and *pitch wheel change*. The status byte of each channel voice message indicates both the nature of the message as well as the specific channel it solely intends to address. The data bytes then provide supplementary details to accompany the message. Each of these is summarized below.

- A *note on* message describes the onset of a MIDI note. Two data bytes are required: one for the note number (ranging from 0–127, with note number 60 representing middle C) and one for the *attack velocity* with which the note is depressed.
- A *note off* message conversely describes the release of a depressed key. Two data bytes are required: one for the note number to be released, and one for the *release velocity*, or speed with which the note is released. (Some synthesizers may alter the decay of a sound, for example, based on the release velocity; others just ignore it.) Note that the sound may still continue: The note off message simply describes when a key is physically released.
- *Polyphonic key pressure* transmits the individual pressure variations of a specific MIDI note after it has been depressed. The first data byte represents the note number, while the second indicates a pressure value. Although not every instrument transmits or responds to this message, it can be used creatively with software synthesizers on the DAW. For example, clavichord-style *Bebung*, a

note-specific vibrato from the German word for "trembling," can be implemented using polyphonic key pressure.

■ *Channel pressure* ("aftertouch") is identical to polyphonic key pressure in intent, except that it represents the global pressure of all the keys played on a specific channel. Only one data byte is required, which indicates the channel pressure value. (Recall that the channel to which the pressure is applied is contained in the status byte.)

■ A *program change* message instructs the recipient to change the "sound" (or more accurately, the "program" or "patch") to which it is currently set on a certain MIDI channel. Only one data byte is required, which indicates the desired program number.

■ *Control change* messages indicate a continuous change in a particular MIDI control parameter on the given channel. For example, using control change messages, the current position of a knob on a synthesizer or control surface can be continually transmitted to a DAW. Two data bytes are required: the specific controller number to change, and the value of the controller. Many different controllers are available, from the global "volume" of a channel to the status of a piano's damper pedal to the spatial location the channel. The official listing of MIDI controller numbers is given in Table 9.1.

■ A *pitch wheel change* message is issued whenever the pitch-bend wheel on a synthesizer is moved. The current value that represents the location of the wheel is sent as a 14-bit message to increase the number of available quantization levels, and so two data bytes are required: one to transmit the least-significant byte, and one to transmit the most-significant byte. It is up to the receiving device to interpret the frequency range of the transmitted numbers, although the median value of the entire 14-bit range denotes the center (equilibrium) position of the pitch wheel.

Channel mode messages. Channel mode messages are used to indicate one of several specific modes in which the specified channel will operate. These messages are actually sent as control change messages using the last seven controller numbers (121–127) available (see Table 9.1). The seven possible channel mode messages are summarized on the following pages.

■ *Reset all controllers*. This restores all controllers to their default positions, whatever the given MIDI device defines "default" to be.

- *Local control on/off.* When local control is turned off, the MIDI device cannot be "played" directly; rather, it only responds to MIDI data. Local control on allows the device to be played as well as to respond to MIDI data.
- *All notes off.* This probably does not need any explanation!
- *Omni mode off.* In this mode, the MIDI device only responds to one channel, whatever that channel is specified to be.
- *Omni mode on.* When omni mode is on, the device responds to messages on all MIDI channels.
- *Mono mode on.* When mono mode is turned on, the MIDI device is forced to respond monophonically. For a synthesizer, this means it can only produce one note as a time.
- *Poly mode on.* On the other hand, turning on poly mode cancels mono mode, allowing the device to respond polyphonically.

System messages. In distinction from channel voice and channel mode messages, system messages are designed to address all connected MIDI devices globally. They are divided into two types: system common messages and system real-time messages.

System common messages include system exclusive, song position pointer, song select, tune request, and end of exclusive messages, as described below.

- *System exclusive* ("SysEx") messages allow proprietary communications standards to be sent among MIDI devices, embedded within the existing MIDI standard. They are often used to download and backup data from a MIDI device and to upload new program (patch) data, but any kind of data can be communicated. In a sense, SysEx messages are an open-ended message type that allow others to "pick up" where the official MIDI specification "leaves off." At the conclusion of a SysEx message, an end of exclusive message must be sent.
- The *song position pointer* message holds a 14-bit value that represents the number of MIDI beats (defined to occur every six MIDI clock ticks) that have elapsed since playback of a sequence (or "song") has begun.
- The *song select* message specifies a sequence to be triggered for playback.
- *Tune request* instructs all devices in the MIDI chain that can respond to this message to tune themselves using their internal self-tuning routine. This was designed with the idea that MIDI-

TABLE 9.1

Summary of
MIDI Controller
Numbers Used
when Sending
Control-Change
Messages

0 Bank Select (coarse)	**64** Damper ("Sustain") Pedal On/Off
1 Modulation Wheel (coarse)	**65** Portamento On/Off
2 Breath controller (coarse)	**66** Sostenuto Pedal On/Off
3 *Undefined*	**67** Soft Pedal On/Off
4 Foot Pedal (coarse)	**68** Legato Footswitch On/Off
5 Portamento Time (coarse)	**69** Hold 2 Pedal On/Off
6 Data Entry (coarse)	**70** Sound Controller 1 (Sound Variation)
7 Volume (coarse)	**71** Sound Controller 2 (Timbre)
8 Balance (coarse)	**72** Sound Controller 3 (Release Time)
9 *Undefined*	**73** Sound Controller 4 (Attack Time)
10 Pan position (coarse)	**74** Sound Controller 5 (Brightness)
11 Expression (coarse)	**75** Sound Controller 6 (Decay Time)
12 Effect Control 1 (coarse)	**76** Sound Controller 7 (Vibrato Rate)
13 Effect Control 2 (coarse)	**77** Sound Controller 8 (Vibrato Depth)
14 *Undefined* 7	**78** Sound Controller 9 (Vibrato Delay)
15 *Undefined*	**79** Sound Controller 10
16 General-Purpose Controller 1	**80** General-Purpose Controller 5
17 General-Purpose Controller 2	**81** General-Purpose Controller 6
18 General-Purpose Controller 3	**82** General-Purpose Controller 7
19 General-Purpose Controller 4	**83** General-Purpose Controller 8
20 *Undefined*	**84** Portamento Control
21 *Undefined*	**85** *Undefined*
22 *Undefined*	**86** *Undefined*
23 *Undefined*	**87** *Undefined*
24 *Undefined*	**88** *Undefined*
25 *Undefined*	**89** *Undefined*
26 *Undefined*	**90** *Undefined*
27 *Undefined*	**91** Effects 1 Depth
28 *Undefined*	**92** Effects 2 Depth
29 *Undefined*	**93** Effects 3 Depth
30 *Undefined*	**94** Effects 4 Depth
31 *Undefined*	**95** Effects 5 Depth
32 Bank Select (fine)	**96** Data Increment (Data Entry Button +1)
33 Modulation Wheel (fine)	**97** Data Decrement (Data Entry Button –1)
34 Breath controller (fine)	**98** Non-Registered Parameter Number (fine)
36 Foot Pedal (fine)	**99** Non-Registered Parameter Number (coarse)
37 Portamento Time (fine)	**100** Registered Parameter Number (fine)
38 Data Entry (fine)	**101** Registered Parameter Number (coarse)
39 Volume (fine)	**102** *Undefined*
40 Balance (fine)	**103** *Undefined*
41 Fine adjustment for Controller #9	**104** *Undefined*
42 Pan position (fine)	**105** *Undefined*

(continued on next page)

TABLE 9.1		
Summary of MIDI Controller Numbers Used when Sending Control-Change Messages (continued)	**43** Expression (fine)	**106** *Undefined*
	44 Effect Control 1 (fine)	**107** *Undefined*
	45 Effect Control 2 (fine)	**108** *Undefined*
	46 Fine adjustment for Controller #14	**109** *Undefined*
	47 Fine adjustment for Controller #15	**110** *Undefined*
	48 General-Purpose Controller 1 (fine)	**111** *Undefined*
	49 General-Purpose Controller 1 (fine)	**112** *Undefined*
	50 General-Purpose Controller 1 (fine)	**113** *Undefined*
	51 General-Purpose Controller 1 (fine)	**114** *Undefined*
	52 Fine adjustment for Controller #20	**115** *Undefined*
	53 Fine adjustment for Controller #21	**116** *Undefined*
	54 Fine adjustment for Controller #22	**117** *Undefined*
	55 Fine adjustment for Controller #23	**118** *Undefined*
	56 Fine adjustment for Controller #24	**119** *Undefined*
	57 Fine adjustment for Controller #25	**120** All Sound Off
	58 Fine adjustment for Controller #26	**121** Reset All Controllers
	59 Fine adjustment for Controller #27	**122** Local Keyboard Control On/Off
	60 Fine adjustment for Controller #28	**123** All Notes Off
	61 Fine adjustment for Controller #29	**124** Omni Mode Off
	62 Fine adjustment for Controller #30	**125** Omni Mode On
	63 Fine adjustment for Controller #31	**126** Monophonic Mode On

Note: For switches and pedal controllers, ≤ 63 = off, and > 63 = on. Control change numbers 120–126 are classified as channel mode messages that apply to an entire MIDI device rather than to a single MIDI channel.

retrofitted analog synthesizers, which periodically drift out of tune as their analog oscillators meander, could periodically re-tune themselves.

System real-time messages, on the other hand, include instructions that are intended to be sent in real time during performance or playback. They are comprised only of status bytes; no data bytes are required. These messages include timing clock, start, stop, continue, active sensing, and reset.

- A *timing clock* message simply sends high-level synchronization information. If synchronization is used among MIDI devices, this message is sent 24 times per quarter note.
- The *start* message indicates that the given sequence (presumably selected either manually or with a song select message) should begin playback.
- A *stop* message instructs all devices in the MIDI chain to cease playback.

- *Continue* requests that the previously stopped sequence resume playback.
- *Active sensing* is rarely used, and devices need not respond to this message to be considered MIDI compliant. When active sensing is invoked, all MIDI devices expect to receive the active sensing message as a kind of "Are you there?" ping at least every 300 msec. If the message is not received at least that often, the recipients can assume that the connection with the sender has been terminated.
- The *reset* message instructs all devices in the MIDI chain to restore themselves to their default, power-on settings.

Extensions to MIDI

MIDI is a well-understood, widely adopted standard for the high-level exchange of representational and control information among devices in a studio. Indeed, the MIDI standard is remarkable in its near-universal acceptance among electronic music instrument manufacturers. However, it is not without its limitations, including its 31.25 kbaud communication speed (inconceivably slow by modern device communications standards), its sometimes goofy or haphazard controller names (like Controller #71, "Timbre"—to which of the many dimensions of timbre does this refer?), and its keyboard-based bias. To accommodate this ubiquity while addressing some of the standard's inherent limitations, however, several companies have developed their own proprietary MIDI-superset specifications and extensions to the standard protocol. These include MIDI timecode (MTC), MIDI machine control (MMC), MIDI show control (MSC), general MIDI (GM), general MIDI 2 (GM2), Yamaha's XS and XF standards, Roland's GS standard, downloadable sounds (DLS), XMIDI (extended MIDI), fuzzy MIDI, and a host of others.

MIDI timecode (MTC) represents at its core the implementation of the SMPTE timecode synchronization standard over standard MIDI connections. (In fact, MTC-SMPTE converter boxes are available to accommodate both standards within the same studio.) As mentioned earlier, it has now been adopted into the MIDI standard as a system message. Using MTC, various devices in a DAW-based studio, like a digital video recorder and the DAW itself, can be synchronized for playback and recording. Pressing the spacebar on a DAW, for example, may trigger playback of a mix while simultaneously invoking the "record" button on a modular digital multitrack recorder. Most DAW multitrack software includes MTC implementation, making integration with other hardware devices much easier using standard MIDI connections.

MIDI show control (MSC) allows theatrical and multimedia devices of all kinds to be controlled using the original MIDI specification and hardware implementation. MSC is widely used in multimedia productions requiring synchronized cueing and control of lighting and staging effects devices, including everything from strobe lights to projectors to fog and smoke.

General MIDI (GM) was proposed to standardize the general layout of MIDI program numbers so that, for example, Program #1 always corresponds to "acoustic grand piano" on all devices, Program #41 always corresponds to "violin," and so on. This standard is particularly useful in gaming and multimedia applications (like Web sites) that want to ensure a modicum of sonic predictability for the viewer of such content. General MIDI-compliant devices also require that percussion sounds always reside on a *key map* on MIDI Channel 10. This means that each note number on Channel 10 corresponds to a different percussion sound. The general MIDI program numbers are provided below in Table 9.2, and the general MIDI percussion key map is shown in Table 9.3.

A backwards-compatible extension to General MIDI called General MIDI 2 (GM2) is sponsored by the MIDI Manufacturers Association to provide additional program names and control parameters. Some instrument makers have instead created their own proprietary extensions to the GM standard with similar goals in mind. These include Roland's GS standard and Yamaha's XS and XF standards.

Some extensions to the original MIDI specification are tailored for specific kinds of devices. Examples of this are found in the GM-Lite (GML) and SP-MIDI standards, which were created primarily for use in mobile devices like cell phones. (That way, you can download new ring tones!) According to the MIDI Manufacturers Association, GML is "intended for equipment that does not have the capability to support the full feature set defined in General MIDI 1.0, on the assumption that the reduced performance may be acceptable (and even required) in some mobile applications."

The scalable polyphony MIDI specification (SP-MIDI), is an alternative standard that allows different MIDI messages to be sent depending on the capabilities of the performing device. For example, a fully GM-compliant device may be instructed to perform a full 16-channel sequence, while another device, such as a cell phone or per-

TABLE 9.2

General MIDI
Program
Numbers

Piano	Reed
1. Acoustic Grand Piano	**65.** Soprano Sax
2. Bright Acoustic Piano	**66.** Alto Sax
3. Electric Grand Piano	**67.** Tenor Sax
4. Honky-tonk Piano	**68.** Baritone Sax
5. Electric Piano 1	**69.** Oboe
6. Electric Piano 2	**70.** English Horn
7. Harpsichord	**71.** Bassoon
8. Clavi	**72.** Clarinet
Chromatic Percussion	**Pipe**
9. Celesta	**73.** Piccolo
10. Glockenspiel	**74.** Flute
11. Music Box	**75.** Recorder
12. Vibraphone	**76.** Pan Flute
13. Marimba	**77.** Blown Bottle
14. Xylophone	**78.** Shakuhachi
15. Tubular Bells	**79.** Whistle
16. Dulcimer	**80.** Ocarina
Organs	**Synth Lead**
17. Drawbar	**81.** Lead 1 (square)
18. Percussive Organ	**82.** Lead 2 (sawtooth)
19. Rock Organ	**83.** Lead 3 (calliope)
20. Church Organ	**84.** Lead 4 (chiff)
21. Reed Organ	**85.** Lead 5 (charang)
22. Accordion	**86.** Lead 6 (voice)
23. Harmonica	**87.** Lead 7 (fifths)
24. Tango Accordion	**88.** Lead 8 (bass + lead)
Guitar	**Synth Pad**
25. Acoustic Guitar (nylon)	**89.** Pad 1 (new age)
26. Acoustic Guitar (steel)	**90.** Pad 2 (warm)
27. Electric Guitar (jazz)	**91.** Pad 3 (polysynth)
28. Electric Guitar (clean)	**92.** Pad 4 (choir)
29. Electric Guitar (muted)	**93.** Pad 5 (bowed)
30. Overdriven Guitar	**94.** Pad 6 (metallic)
31. Distortion Guitar	**95.** Pad 7 (halo)
32. Guitar harmonics	**96.** Pad 8 (sweep)

(continued on next page)

TABLE 9.2

General MIDI
Program
Numbers
(continued)

Bass	Synth Effects
33. Acoustic Bass	**97.** FX 1 (rain)
34. Electric Bass (finger)	**98.** FX 2 (soundtrack)
35. Electric Bass (pick)	**99.** FX 3 (crystal)
36. Fretless Bass	**100.** FX 4 (atmosphere)
37. Slap Bass 1	**101.** FX 5 (brightness)
38. Slap Bass 2	**102.** FX 6 (goblins)
39. Synth Bass 1	**103.** FX 7 (echoes)
40. Synth Bass 2	**104.** FX 8 (sci-fi)

Strings	Ethnic
41. Violin	**105.** Sitar
42. Viola	**106.** Banjo
43. Cello	**107.** Shamisen
44. Contrabass	**108.** Koto
45. Tremolo Strings	**109.** Kalimba
46. Pizzicato Strings	**110.** Bag pipe
47. Orchestral Harp	**111.** Fiddle
48. Timpani	**112.** Shanai

Ensemble	Percussive
49. String Ensemble 1	**113.** Tinkle Bell
50. String Ensemble 2	**114.** Agogo
51. SynthStrings 1	**115.** Steel Drums
52. SynthStrings 2	**116.** Woodblock
53. Choir Aahs	**117.** Taiko Drum
54. Voice Oohs	**118.** Melodic Tom
55. Synth Voice	**119.** Synth Drum
56. Orchestra Hit	**120.** Reverse Cymbal

Brass	Sound Effects
57. Trumpet	**121.** Guitar Fret Noise
58. Trombone	**122.** Breath Noise
59. Tuba	**123.** Seashore
60. Muted Trumpet	**124.** Bird Tweet
61. French Horn	**125.** Telephone Ring
62. Brass Section	**126.** Helicopter
63. SynthBrass 1	**127.** Applause
64. SynthBrass 2	**128.** Gunshot

TABLE 9.3

General MIDI
Percussion Key
Map for General
MIDI Channel
#10

Note #	Instrument	Note #	Instrument
35	Acoustic Bass Drum	59	Ride Cymbal 2
36	Bass Drum 1	60	Hi Bongo
37	Side Stick	61	Low Bongo
38	Acoustic Snare	62	Mute Hi Conga
39	Hand Clap	63	Open Hi Conga
40	Electric Snare	64	Low Conga
41	Low Floor Tom	65	High Timbale
42	Closed Hi-Hat	66	Low Timbale
43	High Floor Tom	67	High Agogo
44	Pedal Hi-Hat	68	Low Agogo
45	Low Tom	69	Cabasa
46	Open Hi-Hat	70	Maracas
47	Low-Mid Tom	71	Short Whistle
48	Hi-Mid Tom	72	Long Whistle
49	Crash Cymbal 1	73	Short Guiro
50	High Tom	74	Long Guiro
51	Ride Cymbal 1	75	Claves
52	Chinese Cymbal	76	Hi Wood Block
53	Ride Bell	77	Low Wood Block
54	Tambourine	78	Mute Cuica
55	Splash Cymbal	79	Open Cuica
56	Cowbell	80	Mute Triangle
57	Crash Cymbal 2	81	Open Triangle
58	Vibraslap	82	Ride Cymbal 2

sonal digital assistant, may be only given a subset of this information to play owing to its more limited polyphonic capabilities.

The ownloadable Sounds (DLS) Standard, introduced in 1997, augments the program sets available on a given MIDI device by allowing digital audio samples to be sent over a MIDI connection and stored inside the device. More robust than the original MIDI Sample Dump Standard (SDS) standard proposed in 1986, the DLS Standard is built around a file format, based on Microsoft's RIFF file format, that allows chunks of standard Wave files to be literally downloaded onto a compatible synthesizer or sound card. Other extensions, like XMF (eXtensible Music Format, based on the more general XML multimedia standard), XMIDI (eXtended MIDI), Fuzzy MIDI, and quite a few others have also been proposed by various groups to address some of the inherent limitations of the MIDI standard.

MIDI Controllers

Because MIDI was originally intended as a general-purpose control standard whereby one device instructs another what, when, and how to do something, it has been co-opted in many ways to create instruments other than the standard keyboard synthesizer controller. MIDI instruments of all kinds—both commercially produced and home-made—can be found, from MIDI guitars to wind controllers and drum kits. In fact, the MIDI control standard is commonly found in DAW control surfaces that emulate the look and feel of a traditional mixing console, allowing intuitive and physical control of multitrack editor-sequencers and other DAW software.

The MIDI specification is also commonly implemented in so-called *alternate controllers* that attempt to meld sensors, microprocessors, and gadgets of all kinds into novel musical interfaces. If a physical quantity can be measured with a sensor (temperature, humidity, ambient light level, etc.), it can be converted into MIDI messages. One of the more exciting areas within modern music technology is centered around examining the myriad possibilities for human control of musical machines, for example mapping gestural movements of performers and dancers into musical control information, and the MIDI standard has indeed played an important role in this area owing to its ubiquity, simplicity, and flexibility (Figure 9.2).

Figure 9.2
MIDI controllers can be made from just about anything. Shown here is the Cook-Leider SqueezeVox controller, an accordion-based instrument for controlling vocal synthesis models.

Beyond MIDI

Despite MIDI "supersets" like general MIDI, XS, XF, GS, and so on, several musical representation and control standards have been proposed that deviate significantly or completely from the MIDI standard. These proposed standards do not attempt to work within the confines of MIDI; rather, they start from scratch with new ways of thinking about musical representational and control issues.

One of the first "post-MIDI" standards proposed was called ZIPI, introduced in 1994 (see http://cnmat.cnmat.berkeley.edu/ZIPI). Although since superseded by other faster and more robust musical communications protocols, ZIPI was unique in its critical thinking about some of the deeper aspects involved in musical control, for example, by including a standard for sending and receiving control information over a network with specific latency restraints. The technology was pursued by Gibson, but it was lost in a string of legal battles between the company and one of the technology's core inventors.

A new control protocol that has emerged recently and been adopted in quite a few DAW software programs is Open Sound Control (OSC), developed by Matt Wright in 2002 at the University of California at Berkeley (http://www.cnmat.berkeley.edu/OpenSoundControl). OSC is a multimedia control standard specifically tailored for network-based communications. Using OSC, for example, a performer at one node of a network (for example at a concert hall in New York) can control the synthesis and performance of sounds at another location (like a computer sitting in another concert hall in Miami).

Many other standards for representations of musical control signals exist. The SKINI message format, created by Perry Cook, is unique among many musical control standards in that it specifies events in plain ASCII text, making SKINI files easily readable and editable, although it is currently only used in the Synthesis ToolKit (STK), a set of C++ routines for synthesizing and processing sound. Other control standards, like the Sound Description Interchange Format (SDIF) and the Structured Audio Orchestra Language (SAOL, created by Eric Scheirer and Barry Vercoe and embedded into the new MPEG-4 specification) blur the line between control and synthesis of new sounds.

Software Synthesis Environments

As discussed in Chapter 2, the genesis of the DAW as we now know it was very much rooted in attempts to use computers to create—that is, to synthesize from scratch—any conceivable sound. Indeed, the ability to synthesize sounds has long been one of the primary draws of using computers in the creation of music, and it is out of this long-standing tradition that many modern DAW software programs have developed.

But specifying the exact sample values in time required to produce a given sound is a very daunting task. At a sampling rate of 96 kHz, this of course means that 96,000 numbers would need to be generated each second. And so to reduce this seemingly overwhelming burden, the precise sample values are algorithmically generated using one of the many available families of synthesis techniques.

These synthesis techniques, which have evolved greatly over the years and themselves are the subject of much ongoing research, can be divided into two broad philosophical categories: *spectral modeling methods* and *physical modeling methods*. Spectral modeling methods, which include techniques like amplitude modulation, frequency modulation, granular synthesis, wavetable synthesis, waveshaping synthesis, wave-terrain synthesis, sinusoidal modeling synthesis, and others, use algorithms that attempt to model the characteristic spectra of the intended sound irrespective of its physical method of production. In a sense, spectral modeling methods do not really care how the sound is physically produced in the real world, but just that the mathematical modeling technique used creates a series of time-domain sample values whose spectrum closely resembles that of the intended sound.

Physical modeling techniques, on the other hand, attempt to reduce the physical interactions of the air, cavities, holes, keys, surfaces, and strings of real-world instruments to a set of equations that describe their operation. These equations can then be solved in real time using varying initial conditions (numbers that might represent, for example, how far a guitar string is pulled before being let go and the physical shape of its resonating body), and the solution to the equations then is taken as the actual time-domain samples that are sent to the D/A converter so that we can hear them. If the physics of the vibrating system are modeled correctly without making too many simplifications or assumptions, then the resulting spectrum should sound sufficiently like its real-world counterpart.

But why would we want to use a software sound synthesis environment instead of a dedicated hardware synthesizer? Certainly, synthesizers do not seem to be going away anytime soon, and more hardware models are available each year. And specifying synthesis parameters in software is probably not as musically appealing to most people as is hammering away at a keyboard synthesizer. But software synthesis environments offer several distinct advantages.

First, it is possible to create virtually any conceivable sound using software synthesis. Using a DAW for synthesis opens the door to an unbounded number of synthesis algorithms, removing the traditional limitations of nonexpandable hardware synthesizers. Second, software synthesis holds the potential to accommodate more efficiently the customized interactions among performers and the resulting sound. Any MIDI controller can be patched to control any synthesis parameter, completely at the discretion of the performer. And finally, because software synthesis environments run inside the DAW rather than an isolated hardware box, a framework for robust interaction among synthesis, video, intelligent "accompaniment" and interaction, and just about anything else is provided within one machine.

We have already spoken a bit about software synthesizers that run as plug-ins within a multitrack mixing environment earlier in the book. These programs typically feature engaging sounds and graphical user interfaces, off-the-shelf usability, and wide commercial appeal. Using commercial synthesis plug-ins, for example, it is possible to literally run a number of legacy digital synthesizers in parallel within the DAW (in case you might want a rack of 500 Yamaha DX-7 synthesizers, for example).

Another approach to software synthesis is to embed the ability to synthesize sounds within a more complete programming environment. While programming a computer is certainly not everyone's cup of tea, many of these kinds of environments are relatively easy to begin using, provided you pay your dues by putting in the time required to learn the basics of how they work. Furthermore, integrating a synthesis engine within a broader musical programming environment affords the opportunity for composers to create more elaborate real-time interactions between performer and computer than are possible with most standalone software synthesis programs or plug-ins. For example, using one of these environments, we could program the DAW to trigger a sequence of audio and MIDI data for playback and synthesize a particular sound whenever it perceives that a vocalist sings a middle C.

These kinds of environments used for both software sound synthesis as well as more elaborate machine-performer interaction and interfaces can be divided into two categories: graphical environments and text-based environments.

Graphical Synthesis Environments

Perhaps the most well-known graphical synthesis environment in current use, Max/MSP, traces its roots to a program called Patcher by Miller Puckette that was written as a MIDI control processing environment on a Macintosh to allow it to interface with the 4X Synthesizer. Licensed to Opcode (a company that has since been acquired by Gibson), the program was developed and rewritten by David Zicarelli, becoming known as the MIDI processing environment Max (in homage to Max Mathews). In parallel to the development of Max, Puckette continued worked on adding audio-synthesis and processing capabilities to his original program, which was called Max/FTS ("Faster Than Sound") and has evolved into the Java-based jMax. A similar freeware program called pd (http://www.pure-data.org) was also developed beginning in 1997 (see Figure 9.3). Likewise, Zicarelli developed the commercial Max/MSP environment that is in widespread use today.

Max/MSP, available from Zicarelli's company Cycling '74 (http://www.cycling74.com), is programmed by dragging and connecting *objects* to create *patches* (programs). It is a true programming language in that it supports data types and various programmatic constructs found in other high-level computer languages, and its functionality can be extended by writing additional objects called *externals*. Externals are available from everything for following the pitch of an instrument to real-time manipulation of digital video using MIDI and Open Sound Control. A companion program of particular interest to DAW users is found in Pluggo, essentially a plug-in version of Max/MSP. With Pluggo, patches can be run in any of various plug-in architectures.

Other graphical software synthesis environments are also available. Of particular note is the Kyma system, available from Symbolic Sound Corporation (http://www.symbolicsound.com), which integrates its own graphical environment with dedicated hardware DSP accelerators.

Figure 9.3
(a) A simple Pd
patch illustrating
FM synthesis;
(b) a SuperCollider
program
illustrating FM
synthesis.

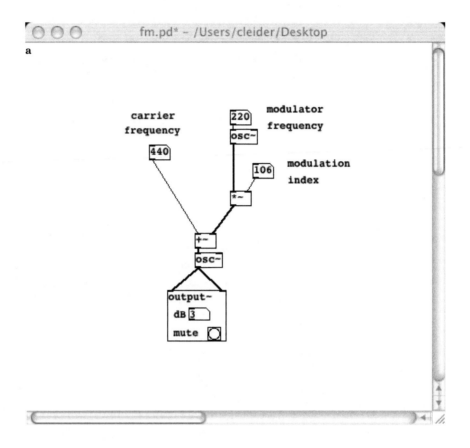

a

b

```
(
SynthDef("fm", {arg bus = 0, carrierFreq = 440, modulatorFreq = 880, index = 5, amp = 0.1;
    var modulator, output;

    modulator = SinOsc.ar(
            modulatorFreq,                      // freq
            0,                                  // phase
            index                               // amp
    );

    output = SinOsc.ar(
            carrierFreq + modulator,    // freq
            0,                                  // phase
            amp                                 // amp
    );

    Out.ar(0, output)

}).load(s);
)


(
Synth("fm", [\bus, 0, \carrierFreq, 333, \modulatorFreq, 33, \index, 77, \amp, 0.15]);
)
```

Text-Based Synthesis Environments

Working with text-based synthesis environments, on the other hand, feels more akin to traditional computer programming. Instruments, sounds, and control data are specified from the ground up, and some text-based synthesis environments are themselves full-blown programming languages. Although in general a bit more difficult and time-consuming to use than off-the-shelf graphical software synthesizers, the payoff can be very big, owing to the increased programmability, flexibility, and expandability.

One of the best-known of these is Csound, which we mentioned in Chapter 2. Although not a programming language, because programmatic constructs like loops are not inherently supported, it has been in use since 1986, and, largely because it has been available ever since for free, it has fostered a large international user base and support community. In fact, a great number of extensions, like graphical user interfaces and even run-time plug-in versions of Csound itself, have been developed. And the Csound environment and related concepts lie at the core of the MPEG-4 structured audio specification.

A number of other "next-generation" synthesis environments are available that integrate the notion of "score" and "orchestra" historically embraced by Csound and its predecessors into one file. This feature is powerful in that the specification of instruments and synthesis parameters, along with their control information ("score") can be placed into a feedback loop in which each can continually modify the others. For example, while running a program in a language like SuperCollider (http://www.audiosynth.com), performance data from a MIDI controller might dynamically modify software synthesis parameters at a fundamental level, while these same synthesis parameters themselves might be used to dynamically create real-time musical notation for the performer to play (Figure 9.3b).

Computer-Assisted Composition and Mixing

Yet another way to expand the DAW's palette is to use it as a compositional assistant to help create or modify musical material. Two broad ways in which the DAW can be used for *computer-assisted com-*

position (also called *algorithmic composition*) lie in the automated generation/modification of control data (for example, notes for a musician or a MIDI synthesizer to play) and in the automated generation/modification of actual audio sample data. The legacy of the former category lies in our long-standing fascination with "intelligent machines" that can emulate the creative processes of humans, while the latter blurs the line between composition and software synthesis. Both represent potentially quite powerful technologies afforded by the modern DAW, but the roots of automatic composing machines long pre-date the digital computer.

Treatises describing the programmed creation of music date back at least a millennium, but the creation of one of the first recorded automatic instruments for composing music is credited to Athanasius Kircher (1601?–1680). In his two-volume book *Musurgia Universalis* (1650), he described the Arca Musarithmica, a mechanical computer that could not only randomly compose music but also compute the date of Easter in any given year and write messages in cryptographic code. Many other devices and musical milestones followed in this tradition with similar goals. Other highlights of this remarkable history include Mozart's composition *Musikalisches Wurfelspiel* of 1787 (the "Musical Dice Game," which he composed by tossing dice to determine the precise ordering of precomposed musical phrases), the computing machines of Charles Babbage (1791–1871) and Augusta Ada King (Ada Lovelace, 1815–1852) in the 1830s, and Joseph Schillinger's 1941 book *The Schillinger System of Musical Composition*, which taught George Gershwin, Glenn Miller, and a generation of others how to compose music algorithmically. The first truly computer-assisted composition, however, is the ILLIAC *String Quartet* (1955), in which Lejaren Hiller and Leonard Isaacson programmed the ILLIAC computer to generate notes for a string quartet to play.

A handful of standalone computer-assisted composition programs are available, but more technically inclined users might find the flexibility of writing their own programs within a high-level computer language, or more likely a software-synthesis language, more appealing. Some algorithmic composition programs are designed to run outside of real time so that musicians can generate and sculpt their output before notating it for instrumentalists or synthesizing the results to disk. Other programs are specifically designed for real-time operation, so that, for example, a performer, perhaps playing a MIDI controller or with the DAW's processing an acoustic signal input via microphone, can interact with a responsive, improvising, intelligent accompanist.

Regardless of the implementation details, however, the methods by which DAWs can be used to compose music can be broadly categorized in three primary ways: according to whether they are designed to run in real time, according to their specific intended use, and according to the algorithmic technique used.

Four primary uses of computer-assisted composition are commonly found: the composer's "assistant," in sonification and auditory display, for automatic stylistic emulation, and for live computer–musician interaction and improvisation. In this first role, algorithmic composition can assist composers and arrangers in many ways: it can suggest new musical possibilities, automatically generate all possible solutions to a specific (often tedious) compositional problem (for example, by producing all possible canonical relationships of two themes, rhythms, etc., for aural evaluation by the composer), or by perhaps generating raw material for a composer/arranger to manually modify and sculpt.

Sonification and auditory display represents yet another area. Here, the goals are often scientific in nature (for example, mapping stock market data to musical notes to hear patterns that might be more difficult to see), or completely musical. Indeed, much great experimental music has been made by algorithmically mapping data from one domain to another. With stylistic emulation, a DAW can be used to automatically analyze existing music and synthesize a similar-sounding music according to some metric or feature set, or perhaps to generate music that represents a mix of features. (Did you ever want to hear Jimi Hendrix meet Beethoven?) And finally, when used in a real-time performance situation, as mentioned earlier, the DAW can be made to intelligently respond in various ways to a performer.

Computer-assisted composition techniques can also be categorized according to the technique used. One option is the use of grammars, automata, and probabilistic models. These methods attempt to model the results of a human composer using mathematical or logical instructions. Artificial neural networks represent a second technique in which the neurologic interactions of the brain are modeled, trained with a data set, and then turned loose to see what happens.

Another technique includes the use of fractals and chaos theory to model musical processes. These can be used to create self-similar musical structures and to iteratively generate musical material. Finally, genetic algorithms represent a new technique in which the interactions of cellular automata are modeled and examined at different stages in their evolution to gather data. For example, the evolution of a "gene pool" of musical elements from a specific composer or musical style might be modeled when "genes" from another composer or style are introduced into the population.

The idea of computer-assisted composition appeals to some and not to others. But considering that it represents just another tool in the DAW's toolbox, it is up to each musician to decide when or if it might be useful for a given musical scenario.

Digital Video

The explosive growth of the use of computers for media creation and modification lies not just in the audio realm, but in the video realm as well. And because the modern DAW can process both digital audio and digital video, new opportunities for their coexistence and co-editing are enabled. In fact, most multitrack editor-sequencers now include support for digital video synchronization and playback alongside a multitrack digital audio mix. And conversely, most computer-based multitrack video editors include quite advanced audio-editing and mixing capabilities. In many ways, the traditional boundaries between audio editing and video editing seem to be blurred: Many of the same compositional techniques can be (and have been) employed in the creation of both.

In general, digital video represents an even higher computational demand than does digital audio. For all but the most compressed and lowest-quality digital video, more memory is required to store each second that passes than for most digital audio files. And as with digital audio files, digital video files can be stored in compressed or uncompressed file formats, although it is common to routinely work with compressed files in most cases, because the playback medium is traditionally quite limited in resolution. (Even the largest new high-definition televisions, for example, still offer nowhere close to the optical resolution of analog 35- or 70-mm film, and compressed video appears just fine on televisions and computer monitors.)

Aside from editing and mixing of video material in a manner analogous to multitrack editing and mixing, the DAW offers two quite unique options over many other hardware-based video editing and processing systems. First, just as software sound synthesis is possible, so too is software video synthesis. Any desired physical object can be synthesized, and it can be made to interact with (or even "drive") the audio synthesis (and vice-versa). Second, the DAW offers unparalleled opportunities for real-time manipulation of digital video, potentially even in response to acoustic stimuli. And many such technologies are enabled by audio software originally developed for the DAW. Consider for example, Jitter from Cycling '74, which is a collection of Max/MSP externals for real-time video processing. Real-time image-processing software can also be used to map live video input (for example, from a camera pointed at a dancer) into musical information.

What's Next?

Speculating on the future potential of the DAW seems like an exercise in futility: Today's cutting-edge audio research might well represent much commonplace technology of the near future. Regardless of said futility, however, I can comment on what seems to represent several of the leading research areas in DAW technology, the fruits of which are already clearly being manifest.

One such area lies in research into digital audio coding systems. To the extent that high-bandwidth communications channels—such as high-definition satellite television, satellite radio, high-definition DVDs, and so on—become more inexpensive and commonplace, coupled with increased consumer demand for multichannel and high-definition audio, so does the need to squeeze out every available bit of that bandwidth in the storage and transmission of this more demanding content. More advanced audio coding takes advantage of ongoing psychoacoustic research from the music cognition community (so that, for example, portions of audio signals that are not physically discernible by the human brain can be swept under the rug, so to speak) making them even more robust while preserving the integrity of the audio and achieving higher compression rates.

Another area lies in the general use of the digital audio workstation as a research and development platform for other kinds of ongoing research, particularly in software sound synthesis technologies, multichannel audio, and the relatively new field of music information retrieval. Thanks to these efforts, respectively, new virtual instruments can be faithfully modeled, modified, and reproduced; sonically immersive environments can be created and sculpted; and vast multimedia databases can be queried for their audio content based on similarity to other audio files or various search criteria.

Finally, the DAW itself serves as a general research platform for the ongoing study and development of new musical instruments, interfaces, and real-time performance capabilities. Wireless electronic musical instruments, novel interfaces that analyze human gesture for its musical intention, audio sound-file servers, and networked musical instruments that communicate not only with each other but with acoustic instruments, human performers, and DAWs have already been developed, and these examples represent perhaps only the tip of the proverbial iceberg in what is musically possible. Thanks to the capabilities of the next-generation Internet, musical telepresence—from virtual remote performances to distributed master classes and concerts—is already taking place and may soon become more commonplace.

All of this activity suggests many interesting possibilities for music of the future. And chief among these must surely be the widespread acceptance of the studio—and thus the DAW in general—as not only a musical instrument itself, but as the progenitor of a music that indeed is afforded by and only possible with technology in general, but with the computer in particular.

For Further Study

Boulanger R, ed. *The Csound Book.* Cambridge, Massachusetts: MIT Press; 2000.

Burns KH. *The History and Development of Algorithms in Music Composition*, 1957–1993. D.A. Dissertation, Ball State University; 1993.

Cook PR. *Real Sound Synthesis for Interactive Applications.* Natick, Massachusetts: AK Peters; 2002.

Dodge C, Jerse TA. *Computer Music: Synthesis, Composition, and Performance*, 2nd ed. New York: Wadsworth Publishing; 1997.

Goodman R, McGrath P. *Editing Digital Video.* New York: McGraw-Hill; 2002.

Huber DM. *The MIDI Manual*, 2nd ed. New York: Focal Press; 1998.

Lehrman P, Tully T. *MIDI for the Professional*, 2nd ed.

MIDI Manufacturer's Association. http://www.midi.org.

Miranda ER. *Composing Music with Computers.* New York: Focal Press; 2001.

Roads C. *The Computer Music Tutorial.* Cambridge, Massachusetts: MIT Press; 1996.

Rowe R. *Machine Musicianship.* Cambridge, Massachusetts: MIT Press; 2001.

Selfridge-Field E. *Beyond MIDI: The Handbook of Musical Codes.* Cambridge, Massachusetts: MIT Press; 1997.

Winkler T. *Composing Interactive Music.* Cambridge, Massachusetts: MIT Press; 2001.

Exercises and Classroom Discussion

1. Elucidate the primary advantages and limitations of the standard MIDI specification.
2. Prepare a short multimedia presentation on a particular alternate MIDI controller.
3. Compare and contrast the features of several commercially available software synthesizers.
4. Compare and contrast the features of two commercially available notation programs.
5. What kinds of musical structures are made possible with algorithmic composition that would otherwise be too difficult to realize?
6. What are some advantages and disadvantages of algorithmic composition?
7. What kinds of sounds are easy to synthesize? What kinds of sounds are difficult to synthesize?

CHAPTER 10

Mastering and Distribution

After creating a final mix on a digital audio workstation (DAW), there is still much work to be done. Specifically, the collection of final mixes intended for distribution on one release must be mastered, duplicated or replicated (we will talk about the difference in a moment), distributed, and marketed. Before the DAW and its accompanying cost-effective duplication solutions, virtually all of these steps in the final stages of assembling a release[1] by necessity took place at a separate commercial facility, most likely using a hired mastering engineer. It is now entirely possible, using professional audio gear at home, to master, copy, distribute, and market a release entirely within your DAW-based project studio. In this final chapter, we address each of these steps in turn, beginning with mastering, proceeding to duplication and replication, distribution, and marketing.

Audio Mastering on the DAW

Mastering is the process of putting the finishing touches, so to speak, on a mix, and it, like just about everything else we have talked about in this book, involves both technical and creative artistry. In general, mastering is not concerned with individual tracks in the mix, and most mastering deals only with the final downmixed tracks of a mix (for example, only the output left and right channels for a stereo mix). As such, most DAW-based mastering can be performed using either a two-track editor or a multitrack editor.

The mastering process typically involves five goals:

- Achieving a particular sound and intangible "feel" to each mix in the release using equalization, compression, and sparing use of other processing algorithms
- Ensuring that the release will maintain its sonic integrity on a wide variety of playback equipment
- Specifically preparing the release, using dithering and other encoding as appropriate, for the target distribution format (whether compact disc, DVD, the Internet, or anything else)
- Ensuring the overall artistic integrity of the release by choosing an aesthetically pleasing order for the individual mixes on the release (the track ordering)

---

[1] Sometimes anachronistically referred to as a "record" or "album," even if the distribution format is not a vinyl record!

■ Maintaining a nominal approximate playback level from track to track within a release

To adequately perform each of these tasks, intimate familiarity with the acoustical characteristics of the studio and equipment used in mastering audio is particularly crucial. There is a niche industry of professionals and professional studios in the business of mastering only; they do not record music, and mastering studios frequently are comprised only of a relatively small control room.

Fortunately, the DAW coupled with a good pair of ears can enable all of this to take place within the computer, potentially saving time and money in producing a release. Let's examine each of the above goals of mastering, beginning with achieving a particular "sound" to a release, specifically with the DAW in mind. Anyone, given quality equipment, good ears, and practice, can learn specifically how to achieve each of these goals, but the first is arguably the most difficult. And professional mastering engineers often advertise their services by listing the artists for whom they have mastered works to give a sense of the kinds of "sounds" they create. Often, we use somewhat nondescript terms like "boomy," "punchy," "tight," "airy," and "transparent," and this intangible quality of a mix is precisely what the "sound" to which we here refer describes.

How is this "sound" created? Clearly, the overall sonic character of a mix, and of course an entire release, depends on a complex interweaving of all the constituent elements—the performers and studio musicians themselves, the microphones and preamplifiers used when recording, the processing plug-ins used, and so on. Once we reach the mastering stage, however, the single biggest contributor to the "sound" of a mix is the global equalization of the entire mix. Experimentation on your own rather than any words I could write here is a much better teacher, but do keep in mind the Fletcher-Munson equal-loudness contours, and the frequency-adjective pairs listed in Table 10.1 may be helpful. Note that the precise frequency region of interest can vary from mix to mix depending on the specific sonic architecture, but the terms in Table 10.1 can be used as a guide to begin.

Also note that combinations of key frequency bands can be addressed to create other specific kinds of "sounds." For example, a sound often described as "tight" and "crisp" can be achieved usually by using equalization and compression to ensure quick low-frequency transient response while emphasizing extreme high and low frequencies and attenuating the midrange.

TABLE 10.1

Several Key Frequency Bands and Adjectives Often Associated with a Peak or Notch at Each Band

<50 Hz	"Heavy"
125 Hz	"Boomy"
200 Hz	"Bassy"
250–500 Hz	"Boxy"
300–1000 Hz	"Nasal"
3 kHz	"Hard"
5 kHz	"Present"
10 kHz	"Etched"

To achieve the second goal of ensuring the compatibility of the mix on a variety of playback equipment, the same techniques mentioned earlier are used. Namely, the mix should be monitored on a number of different loudspeaker systems. This too is a reason why some people use a professional mastering engineer, who already owns a variety of speakers and can offer a set of impartial, "fresh" ears.

Aside from simply monitoring on multiple loudspeakers, dynamics processing is also a key component of most mastering sessions. Compression is usually used to increase the perceived loudness level (by raising the RMS amplitude of the mix), and other psychoacoustically based processing plug-ins are available for this purpose on the DAW. Keep in mind, though, that the degree of dynamics processing is of course related to the target audience as well: Popular music that will be primarily heard in a car will necessarily require greater compression ratios than classical music or electronic music that will probably be heard in a concert hall or at home and for which a greater dynamic range is desired by the audience.

The third goal, preparing the release for copy on the target distribution medium, involves two steps: dithering and encoding. We spoke about dithering earlier; you will recall that this is simply a small amount of noise that is added to increase the perceptual dynamic range of a mix—to let you "hear around" the quantization levels. Dithering at this stage must take into account the target distribution medium. For example, if a work has been recorded and mixed using 24-bit quantization, and it is intended for release on a 16-bit compact disc (actually quite a common scenario today), then the 24-bit master must be bit-reduced using 16-bit dither. Simply converting 24-bit quantization into 16-bit quantization will not sound nearly as good as using 16-bit dither.

The mix must then usually be encoded specifically for its target distribution format. To begin, the mastered work must be bounced to disk, which causes all plug-in effects to be invoked during playback and the resulting output audio stored to a new sound file. For mastering to a compact disc, this file is most often simply a WAV or AIFF file already, so no additional encoding is necessary. If, however, the mix is intended for a DVD-Video release, then it must be *encoded* into the Dolby Digital (and optionally DTS) format. Or consider a mix intended for Internet distribution: You will likely want to encode it in a compressed format, like MP3 or Ogg Vorbis. We will talk more about specific encoding options when we address distribution formats.

The fourth goal, that of choosing an appropriate ordering for the mixes that combine to make a release, is of course a creative choice and matter of personal taste. Various techniques are often employed depending on musical style, for example alternation ("fast" mixes next to "slow" ones, or louder mixes followed by softer ones) and grouping (lumping similar-sounding "sets" of mixes together). Any ordering is possible: Consider that a release with ten mixes can be ordered in 10! = 3,628,800 different ways! Another important creative mastering option lies in the specific inter-track pause time, which can be set to a constant time (like 2 or 3 seconds), or can vary depending on the musical nature of the previous track. Consider, for example, including longer pauses after more introspective tracks.

Finally, the mastering process must ensure that each mix on a release works musically not only in its respective placement within the overall order of tracks on the release, but also that a consistency of approximate dynamic level is maintained. A release in which tracks "peak" at a markedly different level generally will not sound very professional; a constant peak level for each mix on the release is generally much more desirable. Note that levels at this stage should not be adjusted by blindly normalizing each mix upward to a desired peak level; this inherently raises the noise floor. Instead, consider taking several steps back in the whole process, adjusting the overall gain of the mix before dithering and encoding.

Some DAW plug-ins called *finalizers* are specifically made to assist in the mastering process and encapsulate many of the functions we have talked about in this section. Many finalizers employ parametric controls to proprietary internal algorithms, and as such they often feature knobs with such labels as "warmth" or "bite." Hardware finalizers have long been available, but their DAW counterparts can cost an order of magnitude less!

Physical Music Distribution

Music can be distributed using either physical distribution media (like compact discs), or it can be distributed strictly over a network (like the Internet). Some releases are distributed using both paradigms, while some use network distribution as a "teaser" for the physical media release. Others use high-bandwidth value-added content (like games, data files, and so on) enabled by the capacity and immediacy of physical media to discourage piracy. Regardless of the reason, each of these distribution types has its own advantages and disadvantages—merits and flaws that indeed lie in perpetual flux given the current state of the commercial music industry, online piracy, and intellectual property laws.

If a release is intended for physical distribution, several choices are available; compact disc (CD), DVD-Video, DVD-Audio, and Super Audio Compact Disc (SACD) are the most common at the time of this writing. Each medium dictates that its audio content be encoded in a particular format, and it is important to know the capabilities of each medium before preparing for actual distribution of the release.

CD

Compact disc is still the most popular physical distribution format for music today. The standard "Red Book" specification for compact-disc digital audio, developed by Sony and Philips in 1980, dictates that audio be encoded at a sampling rate of 44.1 kHz at 16 bits. (Other compact-disc specifications are also standard, for example the "Yellow Book" standard, which defines the CD-ROM data format.)

A conforming CD is limited to 99 tracks of audio, and each track must last a minimum of 4 seconds. The audio stored must be encoded in PCM stereo (as is the case with WAV and AIFF files), Dolby Surround (for matrixed LCRS mixes), or DTS (for 5.1-channel mixes). Dolby Digital AC-3 encodings are also possible with a hack, but this is not recommended. The maximum amount of stereo PCM audio storable on a Red Book CD is 74 minutes, although 80-minute discs are now possible (but not supported by all players, at least yet).

DVD

DVD, which used to be an acronym for any of various things but now just stands for nothing more than "DVD," provides two formats for distribution: DVD-Video and DVD-Audio. Both of these can store mono, stereo, and surround audio. The key differences are that DVD-Video can store MPEG-2 video, while DVD-Audio can store higher-resolution surround audio at the expense of only being able to play still images during audio playback. DVD-Audio players are currently more expensive than their DVD-Video counterparts, which command a greater market presence. While most DVD-Video players cannot play DVD-Audio discs, the converse is not necessarily true: Most DVD-Audio players are in fact able to play both video and audio discs.

Both of these formats offer key advantages over CD: They are better able to store surround audio mixes in a broader variety of formats, they support higher-resolution audio, and they can store a much greater amount of data. To top it off, the replication price is not necessarily that much higher disc-for-disc than compact disc replication.

The amount of data that can be stored on a DVD depends on whether both sides of the disc are used and the number of layers used (two are available on each side). The capacities of each of these possible combinations in given below in Table 10.2.

TABLE 10.2

Data Capacities of Standard DVDs

DVD Format Name	Specifications	Data Capacity
DVD-5	Single-Sided, Single Layer	4.7 Gb
DVD-9	Single-Sided, Dual Layer	8.5 Gb
DVD-10	Dual-Sided, Single Layer	8.4 Gb
DVD-18	Dual-Sided, Dual Layer	17 Gb

DVD-Video. The official DVD-Video specification mandates that audio must be encoded in at least one of the following three formats:

- *PCM:* From one to eight channels of uncompressed linear PCM (LPCM) audio (e.g., like WAV or AIFF files)
- *Dolby Digital AC-3:* From one to 5.1 channels
- *MPEG-2 Audio:* From one to 5.1 channels and 7.1 channels

Note that Dolby Digital AC-3 and MPEG-2 Audio require a corresponding plug-in be used to encode a mix in these formats. All DVD-Video players are required to include built-in decoders for these formats.

Many commercial film DVDs produced today also include a corresponding DTS 5.1 track as well, which can only be decoded and played back on DTS-equipped home-theater amplifiers. Likewise,

Sony's 7.1-channel SDDS standard is also supported, though a dedicated decoder would be required for most players to decode content in this format at the present.

DVD-Audio.   DVD-Audio, on the other hand, requires that audio be stored at least in uncompressed PCM format. The specification supports up to 6 discrete channels of uncompressed PCM audio at sampling rates of 44.1, 48, 88.2, 96, 176.4, and 192 kHz. In addition, bit depths of 16, 20, and 24 bits are supported. Meridian Lossless Packing (MLP) encoding is also supported, which is a lossless compression format that can yield about a 2:1 compression ratio, effectively doubling the amount of playable audio time on a DVD-Audio disc with absolutely no signal degredation. Other encodings are optional for DVD-Audio discs; these include Dolby Digital, MPEG Audio, and DTS.

SACD

Super Audio Compact Disc (SACD), a format again developed by partnership between Sony and Philips, targets the "audiophile" community with Direct Stream Digital (DSD) encoding of audio. Rather than storing signal amplitudes directly as sample values, as PCM does, DSD instead relies on the principle that if the sampling rate is high enough, all we really need to know is whether the signal increases or decreases in incremental value at each sampling instant. DSD operates at 2.8224 MHz—a full 64 times greater than 44.1 kHz—with a bit depth of 1. How can only one bit be enough? Because the sampling rate is so high, DSD encodes an increase in signal value with a logical 1 and a decrease in signal value with a logical 0.

The DSD encoding used on SACDs can supply dynamic range in excess of 120 dB with a frequency response from 0 Hz (DC) to over 100 kHz. Multichannel audio storage is also possible using DSD: In fact, up to six discrete channels can be stored. A lossless compression scheme is also employed on SACDs that yields approximately 2:1 data compression.

SACDs by themselves require a dedicated SACD player, which can also play audio CDs. But SACDs are rooted in the physical technology of the DVD, and as such, they can accommodate up to two layers. Devoting one layer to SACD content and the other layer to Red-Book CD audio content yields a *hybrid* disc that can be played in both SACD and CD players.

Currently, SACD manufacture can be quite cost prohibitive, much as in the early days of the compact disc. And specialized hardware is required at present to encode audio in DSD format. Therefore, far fewer replication facilities can accommodate the encoding and replication of SACDs than can CDs and DVDs.

Duplication and Replication

Once the entire release has been assembled into its final master, it is then ready to be copied (unless of course it is solely intended for Internet distribution). Two options are available for physical distribution: *duplication* and *replication*. Duplication can be performed at home; this is the process typically referred to as "burning" CDs or DVDs, and it is fundamentally different from replication.

Discs can be duplicated using the CD-Recordable (CD-R) or DVD-Recordable (DVD-R) drive inside most DAW computers, and stand-alone multi-disc duplicators are also available for producing larger duplication runs. These machines work by using a laser to alter the reflectivity pattern of the bottom side of the disc. Specifically, the unit's *write laser* is used to heat a reflective layer on the disc; the heat produced by the laser when it is activated is sufficient to darken a small region on the disc—dark enough so that light from the *read laser* cannot reflect off it. These darkened, nonreflective regions denote a 0, and the reflective surfaces untouched by the write laser denote a 1.

While inexpensive and quick for smaller runs, duplication is not the preferred method of copying discs for distribution. Replication, on the other hand, is a more robust process that ensures the resulting discs are playable on all conforming CD and/or DVD players. In this process, microscopic injection molding is used to physically "etch" pits into the plastic of the disc from a glass master. The read laser of a CD or DVD player is then *reflected* or *deflected* depending on whether its beam hits the reflective surface or a pit. Before replicating a disc on a large scale, it is a good idea to request a test press of the disc. This is a one-off replica that you can try yourself on a variety of CD and/or DVD players to verify its content and playability for yourself.

Most replication facilities can accommodate a variety of delivery formats for your master. When soliciting a replication company, it is important to ask the format in which they prefer that you deliver your master to them.

Networked Music Distribution

Music can also be distributed over a network, in which no physical storage medium is required. This can be far less expensive than producing a complete CD package with full-color cover, program notes, shrink-wrapping, and a UPC symbol, but, for the present at least, it does seem to lack some of the cachet offered by physical media with a physical price tag. Of course, all of this may change soon, as audio-distribution Web sites continue to become more popular and appealing to both musicians and music consumers alike.

Perhaps the simplest way to distribute music over a network is to create a Web site on the Internet. This process can be as simple or as complex as you like (depending on the format of the Web site and your specific experience in creating Web sites), or a company can be contracted to create one for you. When distributing music from a Web site, one must consider the variety of bandwidths people use when connecting to the Internet. Users of everything from slow 56k dialup connections to consumer broadband and educational or corporate T1 lines may attempt to download music from the Web site. This necessitates a clear plan of action when posting content for download. Consider offering a low-bandwidth and a high-bandwidth version of each audio file if the file is available for download; likewise, if a streaming file format, like WMA or RealAudio, is used, most users appreciate being able to choose from low- and high-bit-rate options to accommodate their particular connection speed.

Another way to distribute music over a network is to use a peer-to-peer distribution system, such as Kazaa, iMesh, Gnutella, Morpheus, Hotwire, or Limewire. Peer-to-peer systems work by establishing a proprietary multicast network among all users connected to the network. When a specific file is requested, the client program performs the search over each node on the network until matching files are found; when a download is requested, it comes directly from the user's computer that houses that file. Thus, your machine might connect directly to my machine when downloading a file. When using the Internet, by contrast, files are most often downloaded from a centralized server rather than an individual's computer.

Although often used for software and audio piracy, the legitimate and legal distribution of music routinely is used with these technologies. In fact, many "underground" distribution networks have been established for legally trading noncopyrighted music.

Yet another option for networked music distribution is to use an Internet-based music distribution company, such as emusic.com. This model can be particularly appealing for "independent" musicians, because such companies typically offer a 50 percent royalty for each paid download of a musician's audio files.

Copy Protection

Digital audio, by its very nature, calls for a new level of intellectual property considerations. Unlike analog audio representations, which inherently degrade somewhat with each copied generation, digital audio can be copied perfectly bit-for-bit, resulting in copies that are indistinguishable from the original.

While preservation of intellectual property is important in both the physical and networked distribution of music, it becomes a particularly important consideration when dealing with distribution of high-quality audio files over a network. This is because data files are actually being distributed in this model—a literal bit stream that can easily be copied, just as any other file on a computer can be copied. Physical media, at least currently, provide one layer of deterrence to some users in that they must be "ripped" to obtain an actual data file that can be copied with the click of a button, but even this deterrence is being increasingly mitigated by standalone CD and DVD duplicators and freeware ripping software.

One of the first approaches by the recording industry to protect audio material from unlawful copy was the Secure Digital Music Initiative (SDMI). SDMI (http://www.sdmi.org) represented a consortium of over 200 companies involved in various aspects of music technology, and its focus was the development of digital watermarking technologies for copy protection. Audio watermarks embed an inaudible signal within digital audio files that computer software and digital audio devices are then free to interpret in deciding whether to allow copying of the audio content. After four prototype watermarking schemes were proposed, the SDMI offered an open invitation to attempt to circumvent the watermarks; they asked hackers in an open letter to "[a]ttack the proposed technologies. Crack them."

The success of digital watermarking as a copy-protection tool relies on the impermeability of the proposed watermarking technology to hacker attacks. Digital watermarks can take many different forms: every 1000th bit on a sound file may be slightly modulated to encode a copy-protection message, for example, or the signal may be hidden in much more complex and seemingly unpredictable ways. If a watermark can be detected and decoded by a hacker, then this clearly defeats its intended purpose. Furthermore, the integrity of robust watermarking algorithms, the subject of much ongoing research, must be maintained even after portions of the audio file are deleted or modified by a hacker, posing an additional design challenge to designers of digital watermarking technology.

The four proposed SDMI watermarking schemes were cracked by a small team of academic researchers in 2000, shortly after the open-letter invitation. More precisely, the researchers were able to alter the audio files sufficiently to remove the embedded watermark while maintaining an acceptable level of audio integrity as defined by the SDMI consortium. The Recording Industry Association of America (RIAA), in conjunction with the SDMI consortium and Verance Corporation subsequently threatened a lawsuit if the researchers published their method of cracking the watermark at an academic conference, and the rest is history. Despite this interesting early phase of digital audio copyright protection, watermarking technology in general may play an increasingly valuable role in the protection of intellectual property of all kinds stored in digital form.

Marketing

After a set of mixes has been mastered and prepared for distribution, the final step required is to announce the release to as many people as possible through marketing and promotion. This is all predicated on the assumption that (1) music is a commercial product; and (2) the specific kind of music you have made is marketable and sellable. It has been said more than once, for better or worse, that there are two types of music: that which can make a profit, and that which cannot.

While the first point raised here is beyond the scope of this book (how can you really "own" music?), the second is a more pragmatic consideration. Much music, owing to specific cultural and demographic factors, inherently lends itself to commercial consumption in the marketplace. Other kinds of music do not, and that's okay too. Specialized niche distribution channels, such as experimental music labels, are available for noncommercial or "underground" music.

But many more marketing options are clearly available for commercial music. Aside from traditional distribution channels for commercial music, online companies may indeed play an increasing role, not only in the distribution and marketing of audio sound files, but in that of physical media as well. Consider, for example, companies like CD Baby (http://www.cdbaby.com), which promote, market, and distribute music from independent artists. Many other such companies can assist with marketing the music made on your DAW.

So Now What?

In many regards, the overriding point of this book is this: All the technology, all the gadgets, and all the hard disk space in the world cannot make great music. Humans can make great music, and one tool that we can use to do this is the DAW. The technology can greatly help, but music, by its very nature and psychological impetus, is the sole product of our desire to express something in sound.

The DAW provides an increasingly relevant platform for the creation and production of all kinds of music, both in terms of the completely professional results of which it is capable and in terms of the socioeconomic factors that seem to point toward DAW-based project studios as viable alternatives to traditional high-end professional facilities. To the very same extent that the DAW increasingly permeates all aspects of music creation, new avenues of music production—and certainly new ways of conceiving of music and mixing sound—will continue to inspire, awe, and affect the very tapestry of our musical culture.

For Further Study

Craver SA, et al. "Reading Between the Lines: Lessons from the SDMI Challenge." *Proceedings of the 10th USENIX Security Symposium*, 2001; p. 13–17.

Katz B. *Mastering Audio: The Art and Science*. New York: Focal Press; 2002.

Krasilovsky MW, et al. *This Business of Music: The Definitive Guide to the Music Industry*, 9th ed. Watson-Guptill; 2003.

Owsinski B. *The Mastering Engineer's Handbook*. Hal Leonard; 2001.

Passman DS. *All You Need to Know About the Music Business*, 5th ed. Free Press; 2003.

Rosenblatt B. *Digital Rights Management: Business and Technology*. John Wiley & Sons; 2001.

Taylor J. *DVD Demystified*, 3rd ed. New York: McGraw-Hill; 2004.

Exercises and Classroom Discussion

1. Master and finalize a mix using several vastly different settings. (For example, you might make a beefy low-end grungy mix with no dither, a pristine balanced mix with a particular dither setting, and so on.) Compare and contrast the results.

2. Prepare a list of five different Internet companies that are in the business of marketing and distributing independent recordings, including their prices and services offered.

3. Construct a simple experiment to see how many times within 48 hours a sound file, without any marketing or advertising, is downloaded from your Web site. Experiment with the possible effects that the file name (and the link that points to it) might have on the total number of downloads.

4. Illustrate each of the "sounds" listed in Table 10.1 by mastering the same mix using equalization applied to each frequency region in succession.

Index

Note: Boldface numbers indicate illustrations, tables.